GILBERT WHITE
AND HIS RECORDS

A Scientific Biography

Gilbert White and his Records

A Scientific Biography

PAUL G. M. FOSTER

CHRISTOPHER HELM
London

ISBN 0–7470–1003–X

A CIP catalogue record for this book is available
from the British Library

Photoset by Paston Press, Loddon, Norfolk
Printed and bound in Great Britain by
Biddles Ltd, Guildford, Surrey

Contents

LIST OF PLATES vii

PREFACE ix

CHRONOLOGY xii

1 Introduction 1

2 The Early Years: 1720–1750 9
    Childhood and Schooling 9
    University and Ordination 12

3 Years of Indecision: 1751–1753 18

4 The Garden 23

5 *The Garden-Kalendar* 36

6 'Lights & Shades of ye Prospect': 1754–1765 54

7 The Natural Calendar and *Flora Selborniensis* 62

8 On Equal Terms: 1766–1769 77
    Thomas Pennant 80
    Barrington's *Journal* (1767) 84
    Visits and Visitors 87

9 *The Naturalist's Journal* 95
    Contextual Data 96
    Seasonal Indicators 101
    'Miscellaneous Observations' 105
    1774–1775 106
    Accuracy in the *Journal* 110

Contents

10   The Gibraltar Correspondence: An Approach
     to the Study and Writing of Natural History   115
         Specimens   116
         Advice on the Pursuit of Natural History   122

11   Towards *Selborne* (1789), and After   129
         Deciding What to Do: 1770–1774   129
         Distractions: 1775–1788   138
         1789 and After   146

12   *The Natural History of Selborne*   150
         Sequencing and Arrangement   152
         Use of the Letter Form   156

     Notes to Chapters 1–12   160

     Appendix A: The White Family   204

     Appendix B: White's Extended Garden   206

     Appendix C: Development of the Garden
     at Wakes   207

     Appendix D: 'The Invitation to Selborne'   209

     Appendix E: Blue Mist   212

     Appendix F: Pennant's Queries   214

     Appendix G: Four Examples of White's
     Procedures in the Preparation of *Selborne*   217

     BIBLIOGRAPHY   222
     INDEX   233

Plates

1. White's extended garden

2. The Hermitage

3. *Garden-Kalendar*, 30 September–5 October 1765

4. *Calendar of Flora*, 25–28 June 1766

5. *Calendar of Flora*, 14–20 April 1766

6. *The Naturalist's Journal*, 18–24 June 1775

7. White letter, 30 August 1770, to John White

Preface

Gilbert White, alongside Shakespeare, Bunyan and Wordsworth, ranks as one of the most famous names in English literature—yet he wrote only a single book, *The Natural History and Antiquities of Selborne.* This was first published late in 1788 and has subsequently been issued in over 200 editions: it has become the best-loved and most widely-known book about natural history in the whole world. The sources of this appeal, sustained over two centuries and still vital today, are varied and complex, and it is not the purpose of the present volume to seek an explanation—the materials are scarcely available. In comprehending the appeal, however, writers in the past have chosen to pursue the man at home in his village attentive upon nature, or to study the evocation of the pastoral dream that is *Selborne.* But such approaches are partial. What has been neglected is a concurrent exploration of both within the context of a social and scientific study. It is this, based mainly on White's own records, that is offered in the present work.

White's writing practices over more than four decades are enormously rich and, in his beautifully formed handwriting, a delight to read. They include the *Garden-Kalendar* that he began in 1751 and kept for nearly twenty years; the *Flora* of Selborne (a record of purposeful study over a single year, 1766); letters to a brother who, from 1769 to 1772 and at White's instigation, sent to Selborne from the garrison at Gibraltar regular consignments of natural history specimens; and the acclaimed *Journals* of meteorological and other natural events that he completed almost daily from 1 January 1768 up to the final days of his life in 1793. All these various writings, together with some other documents, underpin the eventual *Selborne* and demonstrate White's systematic, if sometimes halting, development.

At one point in this progress towards a public, White remarked in a letter of December 1770 to one of his correspondents, Daines Barrington, that when a man knows he is indebted to others and is uneasy (because of the absence of opportunity to thank them appropriately) then 'it is a sign he has some honour left'. This is my case. Elucidation of White's records and study of their progressive articulation may seem a somewhat narrow, even attenuated, topic, but the experience of working on the subject has been quite different. White's interests ranged widely, and the resources necessary to approach him have been considerable. In consequence my obligations are many, and I am glad now to be able to extend acknowledgement:

to June Chatfield, Curator at the Gilbert Museum in Selborne, for granting access to archival resources and for suggesting many important research locations; to Roy Hall, Secretary of the Selborne Society, for permission to consult manuscripts belonging to the Society; to Gina Douglas, Librarian at the Linnean Society of London, and Ian Lyle, Librarian at the Royal College of Surgeons, for kindly extending research facilities; and to librarians, archivists and others who have responded directly to enquiries and materially forwarded research at the following institutions: Corpus Christi College, Magdalen College, Oriel College, Worcester College, St Mary's Church—all in Oxford; Pembroke College and the Fitzwilliam Museum in Cambridge; County Record Offices in Berkshire, Clwyd, Devon, East Sussex, Hampshire, Leicestershire, Suffolk, Surrey, Warwickshire, West Sussex, Wiltshire; Lambeth Palace Library, Long Ashton Research Station, Royal Horticultural Society, The Middle Temple, The Inner Temple, Office of the Governor (Gibraltar), Wellcome Institute, Hunterian Museum, Society of Antiquaries, Lancing College, National Library of Scotland, Alexander Turnbull Library (New Zealand), John Rylands Library, The University (Southampton), Science Museum, British Museum (Natural History), New York Public Library, Harvard University, Royal Society, Public Record Office (Kew), Bodleian Library, British Library.

For support, tangible and otherwise, over many years I am most grateful to colleagues at West Sussex Institute of Higher Education both past and present, including John Wyatt (Director), Gordon McGregor, Barbara Smith, Trevor Brighton, Jock Johnston, June Kessell, John Vickers, Emlyn Thomas, John Fines, John Saunders, Tony Barnes, Philip Morris, Adriaan van Noorden, Alison Burgess; and for making valuable comments on my writing, in some instances on specific chapters and in others on a draft of the whole volume, I am greatly indebted to Joss Hiller, Margaret Grainger (who taught me not only the names of White's plants but also the syntax of his *Flora*), John Dixon Hunt, and Sarah Markham (whose work on John Loveday has been important for understanding the context of White's interests in several areas unrelated to natural history). Charles and Jonathan Foster, J. K. Burras (Botanic Garden, Oxford), Maurice Burton, Sir Colin Cole (Garter King at Arms), Arnold Cooke (Nature

Conservancy Council), the late Lionel Higgins, Deirdre Le Faye, and Patrick Moore have also made valuable contributions.

To Joe Richards (of Cilmery), Reg Mutter, Alwynne Wheeler, Viscount Barrington (a collateral descendant of Daines Barrington), Peter Weber (of Darmstadt), Jim Sambrook, Sharon Gilbert, Marian Foster and the late Canon Charles Foster, and to my wife, especial gratitude is due: they have each given more than they know.

Chichester, 31 January 1988

Chronology
Chief Events in Gilbert White's Life and Times

DATE	GILBERT WHITE	OTHER EVENTS
1677		Plot: *Natural History of Oxford-shire*
—		
1681	Grandfather became vicar at Selborne	
—		
1691		Ray: *Wisdom of God*
—		
1702		William III d. — Queen Anne enthroned
1703		Great Storm
1704		Blenheim; capture of Gibraltar
1705		Ray d.
1706		
1707		Buffon b.; Linnaeus b.
—		
1713		Treaty of Utrecht; Derham: *Physico-Theology*
1714		Queen Anne d. — George I enthroned
1715		Jacobite Rebellion
1716		
1717		
1718		Ray: *Philosophical Letters*
1719	Father married Anne Holt	
1720	WHITE born — 18 July, at Selborne Vicarage	
—	After living near Guildford and in East Harting, family settles at Wakes	
1727		George I d. — George II enthroned; Hales: *Vegetable Staticks*
1728		
1729		
1730	Plants oak and ash	Thomson: *Seasons*
1731		*Gentleman's Magazine* (— 1907); Cowper b.
1732		
1733		
1734		
1735		Linnaeus: *Systema Naturae*
1736	First known recorded observations	

DATE	GILBERT WHITE	OTHER EVENTS
1737		
1738		
1739	School at Basingstoke; mother d.	
1740	Oriel College, Oxford	
1741	Father lays out 7 acres at Wakes; observes gossamer — 21 September	
1742	At Whitwell, Rutland, for 3 months	
1743	Receives copy of Pope: *Iliad*; BA degree — 30 June	Banks b.
1744	Elected Fellow of Oriel — 30 March	Pope d.
1745		Jacobite Rebellion
1746	Attends to estate of Thomas Holt in Isle of Ely; tour to Gloucestershire; takes MA	Culloden
1747	Made deacon in Christ Church, Oxford — 27 April; curate at Swarraton; buys Miller: *Dictionary*; smallpox	
1748		Thomson d.; ruins of Pompeii discovered
1749	Ordained priest in Spring Garden, London — 11 March	
1750	Tour to Wiltshire and Devon: returns with sea-kale	
1751	Begins *Garden-Kalendar*; tour to Rutland	Gray: *Elegy*
1752	Junior proctor; Dean of Oriel	Gregorian Calendar adopted in Britain
1753	Visits Bristol Hot-Well; curate at Durley	Linnaeus: *Species Plantarum*; Mulso's uncle appointed tutor to future George III
1754	Grows coxcombs; first entry in *Kalendar* for migrant birds — 22 April	Linnaeus: *Flora Anglica*
1755	Second visit to Bristol Hot-Well	Ellis: *Nat. Hist. of Corallines*; Lisbon earthquake
1756		Seven Years War begins
1757	Candidate for provostship at Oriel; plans Hercules statue	
1758	Sets up Hercules; father dies	Swammerdam: *Book of Nature*; Halley's comet; Linnaeus: *Systema Naturae* (10th edn.); Borlase: *Nat. Hist. of Cornwall*
1759	Tour to Rutland for 6 months	Stillingfleet: *Misc. Tracts*; British Museum opened; Voltaire: *Candide*; Wolfe at Quebec; William Collins d.
1760		George II d. — George III enthroned

DATE	GILBERT WHITE	OTHER EVENTS
1761	Completes ha-ha and builds fruit-wall; curate at Faringdon*	Samuel Richardson d.; Pennant: *Brit. Zoology*; Stillingfleet: *Calendar of Flora*
1762		Rousseau: *Emile*; Hudson: *Flora Anglica*
1763	Uncle Charles d.; inherits Wakes; Battie sisters at Selborne	
1764		
1765	Begins formal study of botany	
1766	Completes *Calendar of Flora* (*Flora Selborniensis*)	
1767	Writes to Pennant; reports discovery of harvest mouse	
1768	Begins *Naturalist's Journal*; meets Banks and Solander	Cook begins first voyage; Royal Academy founded
1769	Revises Forster's tr. of Osbeck; receives Sheffield and Skinner at Selborne; writes to Barrington; inspects first specimens from Gibraltar	
1770		Wordsworth b.; Cook at Botany Bay
1771		Walter Scott b.
1772	Brother John returns from Gibraltar	Coleridge b.
1773		Boston Tea Party
1774	First paper read at Royal Society; discusses *Journal* with Barrington	Omai in London; Goldsmith d.
1775		Jane Austen b.
1776	Grimm takes views of Selborne	Adam Smith: *Wealth of Nations*; American Declaration of Independence
1777		
1778		Linnaeus d.; Rousseau d.; Voltaire d.
1779	Measures rainfall	Siege of Gibraltar
1780	Inherits Timothy; builds Bostal; brother John dies	Gordon Riots
1781	Mrs John White arrives at Wakes	Barrington: *Miscellanies*; Uranus discovered; surrender of Cornwallis at Yorktown
1782		
1783		
1784	Sees Blanchard's balloon	Aikin: *Calendar of Nature*; Great Frost

*Before appointment to Faringdon, White was curate at Selborne on several separate occasions, first in 1751, again in 1756–7, and again in 1758–9; he was to be given office there for the fourth time in 1784 and hold it until his death.

DATE	GILBERT WHITE	OTHER EVENTS
1785		
1786		
1787		
1788	*Selborne* published; brother Henry d.	
1789		Fall of the Bastille
1790		
1791	John Mulso d.	
1792		
1793	Dies—26 June, at Wakes	John Clare b.

CHAPTER ONE

Introduction

Prompted by a correspondent who had written to congratulate him on *Selborne*, Gilbert White entered in his *Journal* for 1 August 1790 a list of the most notable trees in his grounds. Brief as the list is (there are only seven items), it tells us a great deal about White himself, about his family, and about the times in which he lived.

In the eighteenth century, timber was not just an essential material for the construction of all modes of transport, for bridges and houses, and for most of the artefacts used daily in the home, but in a rural area such as Selborne of vital importance for heating and cooking. Indeed, in the early pages of White's book, the importance of trees to the community, and to White himself, is given considerable prominence. In the very first letter there is an account of oaks that furnished much naval timber, and of the 'beech, the most lovely of all forest trees'.[1] Further, when we come to the following letter, we find that White devotes it entirely to trees. There are reports of oak timbers used in the renewal of the bridge over the Thames near Hampton Court,[2] the felling of a famed 'Raven-tree' (also an oak),[3] and the survival into and beyond the time of his grandfather of a planted elm that measured 'near eight feet in the diameter' when it was felled, and this at seven feet above the butt. More appealingly perhaps, we are told of a vast oak set at the heart of the village community, on the Plestor (the village green), which was 'surrounded with stone steps, and seats above', but which was overturned 'to the infinite regret' of the people of Selborne in a great tempest in 1703.[4]

White was writing of these occurrences late in his life, in the 1780s, but although some of them took place in his own time the fall of the oak on the Plestor must have become part of family and local legend, the tempest having raged nearly twenty years before White's birth. At that time, White's paternal grandfather was vicar of Selborne, and as a service to the neighbourhood he attempted to re-set the great tree. Unfortunately, as White records, the attempt was not a success. But this act, of service to church and village, became part of an oral memorial to his grandfather and, alongside a stone tablet erected in

1

his honour for bequests to his community, offered an example that White himself was to observe with respect.[5]

This same grandfather, also a 'Gilbert White', died in 1727 and his widow moved from the vicarage (which stands on the north of the Plestor, adjacent to the church) to a village house west of the Plestor which looked, at the back, straight into the wooded beech slopes of the Selborne Hanger (see Plate 1). The house was known as 'late Wakes', and here, in a need to establish a new household, came to live her son John (White's father), with his wife and family of six children. Shortly after, perhaps as celebration of a permanent return to Selborne (John White and family having led a somewhat peripatetic existence since marriage in 1719), the eldest child of the new generation, Gilbert White himself, planted two trees, an oak and an ash, in the garden at Wakes. It is to this planting (of 1730) that reference is made in the *Journal* of August 1790, at which time White also noted the girth the two trees had attained in the sixty years of their growing. Moreover, the date is suggestive: it seems just possible that the young White's planting was a tenth-birthday celebration. A second pair of trees in the 1790 list, a fir and a beech, is assigned the year 1751 and they must have been planted when White attained his majority (21 years).

Pleasant as it is to speculate on the reasons for the plantings, White himself explains his interest in 1790. His correspondent, Robert Marsham,[6] had told him the measurements of an oak that he, Marsham, had planted as a young boy. Prompted by this information, White measured his own trees and entered in his *Journal* a list of names, dates and sizes. Somewhat with regret one feels, he found that his own trees could offer no rival to Marsham's. His native soil was, however, to serve him as nobly as Norfolk had served Marsham. At a distance of no more than two miles from Selborne he measured another tree: it was the exact dimension required. White's satisfaction at the discovery was considerable, and he communicated his joy to Marsham in the Virgilian enquiry:

> And can [we men] still doubt, and still forbear,
> To plant, to dig, and cultivate, with care?[7]

In face of the uncertainties of old age, and of his times, an assurance of the future had been found in the natural scene: it was a strength to which his life bore witness.

Images of planting alone or even consideration of the uses of timber give, however, a very partial view of the world in which White lived. Apposite as the notion of a stable, steady growth into the grandeur of spreading maturity may be to the hopes of contemporaries of his own standing and sympathy, the realities of many aspects of eighteenth-century life were somewhat different. At the time of White's birth in 1720, the fervent wish of the dominant politicians was that the internecine civil and religious foment of the preceding 100 years should be forgotten. This wish had been reinforced by the successful conclusion of the War of the Spanish Succession and by the peaceful transition from the last of the Stuarts to the first of the Hanoverians. The prospect before the nation had been prophetically announced the year before the death of Queen Anne by the poet Alexander Pope. Commenting on the Treaty of Utrecht (1713), he saw ahead a 'Time . . . when free as Seas or Wind/Unbounded *Thames* shall flow for all Mankind . . . And Seas [alone] join the Regions'—which had formerly been divided by envy and religion. This prospect, of a golden age reminiscent of the greatest days of the Roman Empire under Augustus, in which

> Safe on [England's] Shore each unmolested Swain
> Shall tend the Flocks, or reap the bearded Grain

was encouraged by the stable government of Robert Walpole, and by a commitment to economic prosperity. So successful was the period that, at about the time the White family came to settle at Wakes, another commentator, but one who spoke of the country's secure accomplishments, was able to survey 'the blessings of our isle' and to exclaim that 'Her arts [were] triumphant . . . Her public wounds bound up . . . Her commerce spreading sails in ev'ry sky'.[8]

Yet, if this was the spirit of the times in which White spent his childhood, how different were the days of his late maturity and old age. True, as a man nearing forty he was to witness the apotheosis of Pope's vision in the achievements of Pitt and in the victories of the Seven Years War—of Clive, of Boscawen and Hawke, and of Wolfe. But the military glories of Pitt's strategy which included the security of British trade—more especially of fur, fish, and naval supplies in North America, sugar from the West Indies, gum (and slaves) from West Africa, and spice and tea from the East—were to be short-lived. At the subsequent political peace France was to regain many rights, and in the following decades to claim many more. The American colonies were to assert independence (1783), and the horrors of the Civil War and regicide of the last century in England were to return in revolution just across the Channel. The year of publication of *Selborne* was to witness the fall of the Bastille, and White's decline and death were to occur in the same year as Louis XVI was swept to the guillotine.

To the quiet English disposition, national politics and events of such a high order must seem remote from the gardens and woods of White's beloved Hampshire.[9] Some commentators, especially those who have rested their case on a reading of the *Natural History* alone, have variously praised or bewailed White's absorption in the supposed simplicities of a pastoral year that, seemingly, so inoculated him against the affairs of state. The evidence shows, however, that he was acquainted with such matters as well as, if not better than, most country clergy of his time. Consider the record. On several occasions in the week, national newspapers arrived at Wakes, and from them he frequently extracted news of military disaster or national triumph;[10] as a man of 36 years, he left his bed at 2 o'clock in the morning to witness a review of troops;[11] a lifelong friend, James Gibson, sent him graphic accounts of events during Wolfe's heroic capture of Quebec;[12] a brother, John, spent over 15 years as a chaplain at the garrison of Gibraltar;[13] he bought books that reflected interest in military affairs;[14] and, on returning in 1778 from a visit to a relative who lived three days' riding east of Selborne and hence much nearer France, he was to find a letter enquiring whether he could 'sleep without dreaming of ye French' (that nation having recently concluded a treaty with America, which boded ill for the English colonies, and, in the mind of his correspondent, threatened the Channel coast).[15] Such a picture as these facts give suggests a lively awareness of the world beyond the bounds of parish and church. Try as his friend John Mulso might, to propose that in the face of threatening events we should 'Wrap up [our] Content in the Conclusion of Voltaire's Candide—il faut cultiver notre Jardin',[16] such an easy remedy was to the taste of his correspondent alone. For White himself, something much more active was needed, and in order to sense the context for that activity we need to remind ourselves of some of the ways in which life in eighteenth-century England was different from anything known since. Prominent here are two things: the enormous gap between rich and poor, and the sheer brutality of the period.

In itself the difference between the economic status of rich and poor is striking enough. On the one hand we find White writing that a staple food such as potatoes was grown in Selborne only from mid-century;[17] yet, on the other hand, the *Annual Register* (1770) can report such prodigality at the coming-of-age celebrations of a Welsh nobleman (15,000 guests consuming 18,000 eggs, 200 beasts, 60 barrels of pickled oysters, over 900 assorted birds, 1,300 various fish ...) as to offend belief. Such disproportion in control over resources must not, however, be thought of as an issue in political debate: at least, not until the final decades of the century. On the contrary, instead of inveighing against flagrant inequality, it has to be recognised that society was strictly hierarchical—and unable to be otherwise. Nevertheless, a labourer might receive for work done reward of no more than £10 in a year, and a peer collect from rents £10,000 in the same period; a prosperous county landowner could spend in twelve months more than a humble tradesman might earn in a lifetime; one man could be Lord Lieutenant for three counties simultaneously and hence at the pinnacle of local patronage, and in control of the militia; and voters in some boroughs might not be required to vote for nearly forty years, which for many was more than a lifetime.[18] What all this meant for the average citizen, who was not as the word implies a town-dweller but a peasant farmer,[19] was that, except for Sundays, high days and fair days, expenditure of time and energy was devoted totally to survival. And yet—and here we need an imaginative eye—the requisite survival skills were not those associated with late-twentieth-century urban and state bureaucracy, but a set of physical demands that most of us today would find insupportable. Peat and timber had to be cut for fuel and heat; hay and wheat to be garnered, the latter in competition with geese, fowl, and pigeon;[20] and water fetched from well[21] or stream, to be heated in pots on the open fire. There was a near-absolute dependence on the hours of daylight for conducting the daily round;[22] and even if, say, one's task lay in a contribution to the haulage of timber from the Sussex forests to the naval yard at Chatham (which is thought to have taken up to three years),[23] the boundary of life had, of necessity, to be the parish and its neighbouring parishes; and all this while living in dwellings that for the most part would be unglazed, straw-strewn, and verminous.[24]

A similar picture, similar that is in its difference from today, is found if we turn to levels of cruelty. In England, slavery was not abolished until 1772, at which time there were thought to be nearly 15,000 slaves in the country, many of whom were bought and sold at local fairs and markets. Urban mobs, in the absence of an efficient policing authority, caused ten times as much destruction in London alone in a week in 1780 as occurred in Paris throughout the Revolution. For certain offences, right into the 1780s, women could be burned alive at the stake; and yet, in an age that turned to drink as an anodyne, a woman who put an infant on the fire at a christening, thinking in her stupor the child to be a log, was discharged by magistrates. And until 1777, the heads of Jacobites could be spiked on Temple Bar, as a public reminder of the commitment to the Protestant cause.[25]

In face of such conditions, and such barbarity both in the natural as well as the human scene—one parish killed over 5,000 moles in a single year, another killed 14,000 sparrows in ten, and frogs were skinned to see how long they could survive[26]—it is remarkable that so many worthwhile, even civilised, achievements were accomplished. At one level, it is easy to argue that since life for the ordinary mortal was so bad—in parts of London three out of four children did not survive beyond six years, and adult life expectancy was less than 40 years—men should endeavour with all available human skill and knowledge to alleviate hardship, improve conditions, and reduce cruelty and barbarism. Such an argument has a grain of truth. The growth of hospitals from 1730 onwards was active,[27] and so was the wish to improve cattle, develop efficient drainage schemes, and extend crop yields. In several letters of *Selborne*, White himself makes reference to these latter aspects of what has come to be known as the humanitarian movement, believing for example that 'he would be the best commonwealth's man that could occasion the growth of "*two blades of grass* where *one* alone was seen before" '; and similar praise is assigned to the man who could extend his knowledge into the insect kingdom and thereby reduce the attentions of such creatures on man's crops.[28] Important as these suggestions are, however, they all occur late in *Selborne*, and did not form part of the genuine correspondence with which the book began. That that is so is significant: such hopes and expectations were never part of the fabric of White's intellectual background. For one thing, he was aware of how slow change in the parish would be; for another, the whole climate of his thinking had been formulated in an earlier age, prior to anything that was recognisably part of our modern world. Two examples of that climate must suffice. They concern, firstly, population in its gross numbers; and secondly, the balance in the population of male and female.

Figures for the total population of England at any time during the century are unobtainable, since there was no census until 1801. As the century progressed, there was fierce debate about how such figures as there were (derived in the early years from parish registers) contributed to an argument between parties known as the Ancients and the Moderns. In essence, the dispute concerned the superiority, or otherwise, of the then state of the world compared with the achievements of the classical worlds of Greece and Rome.[29] The Moderns based their argument upon the progress in their own time of discovery, commerce, and thought, several aspects of which were endorsed by White; the Ancients based theirs on the glories of the past, in conjunction with a literal interpretation of the Bible. This last, in relation to human existence, seemed to suggest that the period in which a man was permitted life and breath was declining. For example, before the Flood was Methuselah (who died aged 996), after

the Flood was Abraham (who lived for 275 years), and so on, until
Joseph, of the coat of many colours, who lived for only 110 years. In
later ages it was known that men lived, if they were fortunate, the
allotted span of three score years and ten, and that most people in the
world of the early eighteenth century did not achieve even that. The
conclusion was drawn that Nature was running down; that history,
from the time of Adam and Eve's expulsion from Paradise, was one
long, slow decline into total extinction. To assign any precise time at
which the argument was resolved would be unrealistic, though most
commentators would accept that at the beginning of the century the
general climate of opinion was to look backwards, but that by the end
of our period, in the 1790s, most people were aware of a decided swing
away from the trammels of the past, and faced the future with courage
and fierce convictions. Between these two attitudes there seems to
have existed a period best described as steady state, a period in which
many men whose views were formed by thinkers embracing the past
found a certain contentment with things as they were. Such a man was
White. Not, it must be understood, that there is any trace of self-
satisfaction in his writing or accomplishment; on the contrary, his
work shines with a purposeful determination, however modestly
expressed, to adduce evidence that God's dealings with the world had
been right in the past, and were still right now, if only we could
understand them. Indeed, his scientific endeavour, apart from the
considerable intrinsic enjoyment it afforded him, was pursued
primarily with such an aim in mind: to show, experimentally, the
bounty of Providence.

Illustrating this, in relation to gross population and to its sexual
balance, we can turn to one of the introductory letters of *Selborne*
(Letter 5 to Pennant). There, in tabular form, in a footnote, White
provides a numerical account of the state of the parish 'taken October
4, 1783'. The data given relate to the average of baptisms and burials
for each decade between 1720 and 1779, classified into males and
females, with more detail for each year in the period 1761–80. The
numbers themselves are not striking: the parish population at the
time of the death of his grandfather was 'computed at about 500', and
White himself for 1783 counted the total as 676. What fascinates is the
conclusions White draws from the data. Some of the most general are:

> Baptisms exceed burials by more than one third. . . . a child, born and
> bred in this parish, has an equal chance to live to above forty years.
> Chances for life in men and women appear to be equal.

These conclusions, together with the summative statement for 1761–
80 that 'the births exceeded the deaths . . . 140', demonstrate not only

White's interest in matters of population, but also his satisfaction at what was happening within his native bounds. The parish was not in decline; quite the contrary. The numbers were increasing (Pennant 5 concludes that 'the parish swarms with children'), and male and female had equal chances. What pleasure he must have found in the figures! His mentor, John Ray (1627–1705), in a book that by 1750 had reached a twelfth edition, had reminded his readers of the 'great Design of Providence to maintain and continue every Species', and it was 'to this Purpose' that 'all Creatures should be made Male and Female'. Further, in respect of Man, the 'great Wisdom of the divine Creator' was such that 'Pleasure [was] annex'd to those Actions that are necessary for ... the Continuation and Propagation of the *Species*'.[30] Ray writes here of the survival not merely of Man, but of Nature also. It is in those terms, of the discovery of Divine Providence in such matters, that we should interpret data about the human population of the parish of Selborne, as well as the many references in White's work to 'generation' in the animal kingdom.[31] The world had been created, so those of White's persuasion believed, for Man's good, for his enjoyment even. And the parish of Selborne was ensuring that things should stay like that.

The Early Years: 1720–1750

White was born at Selborne on 18 July 1720. At the time, his parents were staying in the Vicarage, the home of his grandfather, Gilbert White senior. His father, John White, had married Anne Holt the previous September, and White was the first of eleven children, eight of whom survived to adulthood. John's wife had brought with her a modest inheritance and this, coupled with no great success in his profession (that of barrister), persuaded White's father to retire. For a while the newly-married couple were uncertain where to settle. They lived first at Compton, near Guildford, and then briefly at East Harting, a small village near Petersfield. On the death of White senior in 1728,[1] however, they decided to set up house with his widow at Wakes: it was the home White was to grow to love, and where he was to die.[2]

Childhood and Schooling

The factual record of White's first ten years is slight. Although we know of the birth (and death) of brothers and sisters, and of the marriage at about the time the family came to live at Wakes of two aunts on the same day, information about White himself is almost a blank.[3] Writing to a niece in the 1780s, he was to say he had been 'bred up by hand'[4] and, as noted earlier, there is record of tree-planting. But that is about all.

In the following decade the record is a little fuller. One of his recently married aunts died after childbirth; his mother gave birth to two more children, a sister and a brother; and, as a result of a bequest by Gilbert White senior, the metalling of an important route from village to forest was supervised by his father.[5] In addition, a new well was sunk at Wakes—this again overseen by John White—from which, at the time of its first cleaning in 1781, some of the marbles with which White had played as a youngster were recovered.[6]

More relevantly for the concerns of the future, amidst the family events and responsibilities there began to emerge a sense of purposeful development. This is shown in his formal education which he began in the 1730s. Details come primarily from White himself: from his pocket-book, from a reminiscence in *Selborne*, and from an entry in the *Journal*;[7] the precise dating of events, however, is unknown. What seems to have occurred is that after early instruction at home, conducted most probably by his father, he was educated first at Farnham, and later at Basingstoke. What is certain, for the record is in White's hand, is that while he was at Basingstoke his accomplishments were substantial. The books he carried with him in January 1739 numbered over thirty volumes; most of them were classical primers (Latin), but they included several books about Christianity, and an edition of the most considerable (and enduring) recently published poem, James Thomson's *The Seasons*.[8]

How long White spent at Farnham and Basingstoke cannot be established, but it is known that the latter was where he became acquainted with the Wartons: the controversial father, Thomas, and his two sons—Joseph, a future Head Master of Winchester, and Thomas, who was to become Professor of Poetry at Oxford and Poet Laureate. Warton's tuition enabled White to proceed to Oriel College at Oxford, but, the very day he was admitted a Commoner at Oriel, 17 December 1739, his mother died at Selborne in consequence of an attack of measles. This delayed his arrival at university, and he took up residence in college the following term, in April 1740.

If there is such a dearth of factual detail concerning these first two decades of White's life, there is scarcely more to be gleaned about the informing influences of the adults with whom he had spent nearly twenty years. Here, White himself, together with several family portraits, is the main source. The most important figures seem to have been his paternal grandparents, and his father. The tradition of charitable acts and regular improvement to one's estate, established by Gilbert White senior both in his lifetime (he held the cure at Selborne for over 45 years, and accomplished major repairs to church and vicarage) and by bequest (in addition to those mentioned, further monies were left for church use), seems to have passed to his son John. Although an occasional commentator has speculated that John White was something of a disappointment,[9] especially perhaps to his forthright father, his accomplishments were far from mean. He attained professional status, became a barrister at the Middle Temple, was learned in the classics and the practice of the law, and held office as Justice of the Peace; he honoured the parochial interests of his father, and carried forward the tradition of improvements to his dwelling—in 1741, he laid out the gardens at Wakes.[10] Further, in revealing remarks by White and Mulso, it is clear that he inspired affection.

From the latter we know of his musical interests (to be inherited by White) and how, when the Mulsos were on horseback one summer, they were able to 'amble along by Hedges under ye wind like [White's] father';[11] and, when the future career of a nephew was under discussion, White himself wrote to his brother: 'As to law, I have nothing to say about it; lawyers get all the money. Our father, you know, did not approve of it'.[12] Equally illuminating are two notes White made in the *Kalendar*. The one avers that John White was 'a nice observer of [a weather-glass] . . . for full 37 years'; while the other is a record of White planting for his father during the latter's final illness and decline 'some bunches of single snow-drops in bloom under my Father's window'.[13] From these few details, it is possible to see, not someone who was a disappointment to his family, but someone who, in a quiet, gentle, modest way, wished to be different from his father.

White's memory of his paternal grandfather can have been little more than that of a somewhat forbidding elderly clergyman—after all, he was only seven-and-a-half when his grandfather died. His grandmother, however, lived long enough for him to know her well. Rebecca White was to reside at Wakes until her death in her 91st year in January 1755. In contrast to her husband, who had been born and bred in Oxford (and who, on appointment to Selborne after a brief period as vicar at a small parish in Norfolk, resigned his Fellowship of Magdalen College, Oxford), Mrs White was a local girl. The somewhat grand figure of the new incumbent at Selborne, with his even grander background—his father, Sir Sampson White, had been twice mayor of Oxford and knighted by Charles II at the time of the coronation in 1660 for loyalty to Charles I—allied to a girl from the adjacent parish of Newton Valence whose father was a simple yeoman farmer from Nore Hill, makes the marriage seem unusual.[14] Mulso's response to the news that White's grandmother had died is, however, revealing: 'You are very pious in your Character of your Grandmother; may You inherit the same Fullness of Years, with the same Simplicity & Worth of Heart'.[15] Undoubtedly, she was a country woman, with country sympathies and country certainties. They were values White must have imbibed in his childhood and youth almost as strongly as the air of Selborne. He was to find in himself both the learning of father and grandfather, and the wisdom of hanger, beech and stream. The union, unfruitful in children—for White was to live a bachelor,[16] was to call forth the greatest paean of praise known in English prose for the beauty of God's designs for Man, the paradise that is Earth. If Rebecca White had known that, she would have been well pleased.

University and Ordination

The spring of White's arrival at Oxford in 1740 was of famine proportions, with northeast winds so strong and cold that many bird species wintering in southern England departed not at the usual time (in spring), but stayed as late as June. The next months were exceptionally dry and led in autumn to memorable opportunities for the sportsman, especially for those whose game was the partridge.[17] Such a man was White. Just when he acquired the skill of handling a gun is not recorded, but the preparations he made for the sport, and the pleasure he gained from the consequent 'field-diversions', are unquestioned. For the former, account books can be consulted; for the latter, White's own evidence, as well as the reminiscence of a friend, is strong. It seems that he kept meticulous accounts throughout his adulthood, and certainly from the time he proceeded to university.[18] The discipline of maintaining such data over so many decades (in total, at least fifty years) illuminates not merely the social station of the family, but also the personal trait that was to serve him so well in the keeping of records about his observations in natural history. These early records are devoted to routine matters of daily expenditure, but a clear picture emerges of monies devoted to 'field-diversions'; and some comparative costs can be established. For example, shooting expenses concerned exclusively with the instrument of death, that is, with minor repairs to his gun and with the purchase of gun-flints, powder, shot and papers, amass in a year to the equivalent of, say, 20 lb of breakfast sugar, twenty quires of writing paper, or 15 Sunday dinners.[19] In itself, that may not seem very much, but we need to remember that items such as sugar and paper were comparatively much more expensive in White's day than they are now; and that costs in relation to a gun alone must be supplemented by other expenses— hunting dogs had to be bought, fed and attended to,[20] hunting boots to be paid for, and the costs met of hiring a boy for a day's shooting.

The persistence of payments of the kind indicated suggests a regular, even relentless, pursuit of avian game, and confirmation of White's involvement during his Oxford years is given by Mulso. Writing in August 1780 about a dish of snipe, he enquires of White:

> Was our Dish, or was it not, a Curiosity? I do not remember your ever shooting a Snipe at Oxford in Summer, where there used to be Plenty in Winter—

and continues philosophically:

> at that Time You used to practise with your Gun in Summer to steady your Hand for Winter, & inhospitably fetch down our Visitants, the

> Birds of Passage. What You was then is my son John now . . . a complicated Murderer.[21]

In some respects White acknowledges this view, for at about the time Mulso was reminding him of his youth he was composing letters that were to form the prefatory content of *Selborne*. There, writing admittedly about deer-hunting, but with application much more broadly, he comments that the temptation to hunt is scarcely to be resisted 'for most men are sportsmen by constitution . . . [There] is such an inherent spirit for hunting in human nature, as scarce any inhibitions can restrain'.[22] Today, we may find such a view alarming, even distressful, and not helped by entries laconically reported in the *Kalendar*: 'Destroyed 24 bullfinches'—in spite of the explanation being in the cause of gardening endeavours, for the birds had been 'very hard on the Cherry-trees, & plum-trees, & had done a great deal of Mischief'.[23] If accounts such as these disturb, we need to exercise our imaginative eye in directions that are unfamiliar, possibly almost distasteful. Two hundred and more years ago, attitudes to the animal kingdom were very dissimilar from the feelings many people hold now. Whereas late in the twentieth century (and for a good period of time before that) Man is prepared to assign in his thinking a place for the higher creatures commensurate, in some instances, with Man himself, and becomes ever more sensitive to the extinction of species and to the widespread ecological damage caused by pollution of river and ocean, forest and air, in the mid-eighteenth century the spirit of the times was quite different. Then, not only was he still struggling, in face of famine, disease and other natural disasters (the Lisbon earthquake of 1755 reminding people in England of the risks inherent in their own tremors),[24] to ensure a confident survival for the race, but orthodox religion sanctioned an attitude to the Earth that placed the animal kingdom at Man's disposal. White certainly endorsed the standard view that, because of the hierarchies in the natural kingdom, Man was entitled to dispose of its inferior productions in ways that satisfied his own needs.[25]

It would be wrong, though, to see White's activities as mercenary, utilitarian, or merely pleasurable: coeval with his pursuit of game there was born an interest in nature that, allied with the forces of his upbringing, was to yield the justification for the enquiries that were to underpin *Selborne*.

In the autumn of his second year at Oriel, during a vacation at home, he was 'intent on field-diversions [and, rising] before daybreak . . . [he] found the stubbles and clover-grounds matted all over with a thick coat of cobweb . . . [such that, when] the dogs attempted to hunt, their eyes were so blinded and hoodwinked that they could not proceed, but were obliged to lie down and scrape the incumbrances from their faces with their fore-feet'. Continuing the account, White provides what must be the most famous description in English of a shower of gossamer: so complete was the shower that he could claim the 'webs were not single filmy threads . . . but perfect flakes or rags . . . [which] hung in the trees and hedges so thick, that a diligent person sent out might have gathered baskets full'.[26] The phenomenon impressed the young White to such an extent that, when he came to write of the event over thirty years later, he not only assigned a precise date and time to the experience, but was able to convey the marvelling surprise and curiosity with such precision that the scene is as vivid for us as readers now as it was for White at the time.

To say of the event that its power was formative of the future natural historian is to claim too much; and yet, as White returned home, he discovered he was 'musing in [his] mind on the oddness of the occurrence'. In itself the admission may not amount to a great deal, but for it to happen to a young man, just past twenty, who had been reared in freedoms of time and place that we find difficult to imagine, it is striking evidence of the way the experience affected him. At the very least, the phenomenon tightened the springs of enquiry that was to unwind gradually, and persistently, as the years progressed.

Exploits in the field were not the only diversions that engaged White during his Oxford years. He played chess, attended concerts, won and lost at cards; purchased wine by the gallon, cider by the hogshead; regularly visited the coffee-house, and bought an astonishing variety of tea—the accounts list green, hyson, balm, congo and bohea. And he travelled, although not as one might first think on what came to be known as the Grand Tour, for to do that (visit the famed sites of antiquity on the Continent) was probably beyond his means, and certainly outside his temperament and inclination. True, now and then he indulged himself in an excursion—to look at a garden, or a grand house—but the occasions seem to be few, and were probably as much social as cultural in their intention.[27] The tours White did make were more homely: to see friend and relations, or to attend to family matters. In the first few years at Oxford he spent most of his vacations at Selborne, but in the summer of 1742 he stayed for three months with an aunt at Whitwell in Rutland, and four years later visited East Anglia to be resident for six months in the Isle of Ely attending to the estate of a lately deceased relative. And he made several shorter family visits, to Ringmer (near Brighton) and to London.[28]

Amidst these varied diversions and journeys, there were periods of academic learning. He took his BA degree in June 1743, was elected a Fellow of his college in March the following year, and obtained his MA in 1746.[29] The demands of courses of instruction may not have been onerous (mid-eighteenth-century Oxford and Cambridge are often criticised for the absence of serious endeavour),[30] but formal lectures are not the only mode of learning available to a student: friendship and shared talk are also important.

To overhear at this distance of time the conversations White engaged in is impossible, but a glimpse of the world he frequented is available in the form of the series of letters he was to receive for almost fifty years from his college friend, John Mulso, who later was to become parson at a parish (Meonstoke) near to Selborne and a prebendary at Winchester Cathedral. Unfortunately none of White's replies are preserved, but from the way Mulso writes it becomes clear that the Oxford years were in part devoted to extensive reading in the literary field, mention of things to do with science, theology, politics, even of natural history being noticeably infrequent. This bias to the literary must reflect the leanings of the writer, but the allusions to one author after another fall from Mulso's pen in such a fashion that there can be little doubt he is writing within a shared understanding and an interest on which he could rely. Scrutiny of the first fifty or so letters White received from Mulso over a period of ten years reveals familiarity with a galaxy of classical and English literature. That there should be a dozen references to things classical is not surprising (the period is often termed the Augustan period), but that there should also be a rich variety of references to the English tradition is less expected. If Chaucer is omitted on grounds of barbarity (although White will quote him in *Selborne*), many other major authors have a place. Spenser, Donne, Milton, Dryden and Pope are the poets; Shakespeare, with allusions to seven plays, and Otway are the dramatists; Swift, Fielding, and the *Spectator* of Addison and Steale represent prose. But that is not all: Mulso mentions also a number of contemporary authors. The most distinguished are Samuel Richardson, the novelist, whom the Mulsos seem to have met frequently, and the poets Edward Young, William Collins, and the Wartons, both Joseph and Thomas. These last, together with Thomas Gray, were the leading poets of mid-century, taking verse in a new direction after the dominance of Pope. The degree to which Mulso was a participant in the talk and favours of such men cannot be gauged; but we know that White knew the Wartons intimately, that he took tea in Collins' rooms in Oxford,[31] and that Mulso looked to White for poetic encouragement. Responding it seems to an enquiry about his own poetic activities, Mulso wrote deploring his current isolation (he was then in

Kent), and distance from the life of the university:

> I don't find at present any great Propensity to Poetry, & I find that a
> great Deal of ye Inspiration that Poets talk of, is owing to ye Company
> that they are with, & not to ye nine Muses & ye whispers of Apollo; I
> have not seen a verse since I left You, nor hardly heard one quoted, so
> that I am not incited to it.[32]

A more significant indication of White's own inclinations would be difficult to find. Apart from the evident pleasure he derived from conversation about the poets, he also wrote himself. The earliest form this took was undoubtedly formal translation of the classics as part of the requirements of schooling, but the taste for versifying stayed with him. In 1746 he sent Mulso a translation of a Horace ode, and in the following year a poem wholly of his own. It is titled 'The Invitation to Selborne' and was to be his most sustained production.[33] The dominant interest of the poem lies in its matter: the manner is hesitant and conventional, and belongs to the years of an age that was just ending. What White offers is a period evocation, in 'prospect' form, of his native place—its surroundings and distinguishing features, both natural and antiquarian. In the course of the poem, he names his own modest estate as a 'rural, shelter'd, unobserv'd retreat'.

In spite, then, of the diversions of Oxford and the regular visits to friends and relatives, the satisfaction and delight felt within his own parish remained strong; and it is no surprise to find that he sought entry to the church (he was made deacon on 27 April 1747 in Christ Church Cathedral, Oxford, by the Bishop, Thomas Secker), and was appointed curate to his uncle, Charles White, at Bradley and Swarraton, villages near to Selborne.

At this point it would be natural to think that White might have disappeared into the leafy woods of east Hampshire for the remainder of his life, but there are signs in the final years of the decade of a continuing restlessness. Checked by an attack of smallpox in autumn 1747 (which kept him at Oxford for several months and through which he was nursed by an elderly retainer sent from Selborne),[34] he maintained no settled life. At one moment he is visiting Mulso at Sunbury-on-Thames (where the latter now held office) and preaching four sermons; at another moment he is again in Ringmer or staying with a college friend in Gloucestershire;[35] he records expenses in connection with field diversions and buys an important volume about gardening, Philip Miller's *Dictionary*; in 1749 (11 March) he is ordained priest—this time in London at the chapel in Spring Gardens— by the Bishop of Hereford, and in the following year makes a long journey to see friends as far afield as Wiltshire and beyond, to the South Hams of Devon. Although the final months of 1750 find him at Oriel (a visit in part necessary because of difficulties facing his brother John),[36] corroboration that life there was not wholly to his taste is provided by Mulso in a letter of 14 December that year:

> I love the Thoughts of Oxford, yet I agree wth You, it has not the same Charms for me that it had . . . You are soon tired of what You call a Coll: Life . . . [37]

This would seem conclusive enough. But the pull of Oxford was not yet complete, and its resolution—in favour of Selborne—was to be delayed some years.

CHAPTER THREE

Years of Indecision: 1751–1753

The first year of the new decade began with two important events for the future author of *Selborne*. On 6 January 1751 White's little sister, Anne, married Thomas Barker of Lyndon Hall, Rutland, and on the following day, Monday 7 January, White made the first entry of a new diary, which he headed 'The Garden Kalendar for the Year 1751'.

Acquaintance with the Barkers in White's generation derived from an earlier marriage, that between Elizabeth White (a sister of White's father) and John Isaac, sometime Rector at Whitwell, a parish adjacent to Lyndon. This had occurred in 1719, and there seems to have grown from that time onwards the closest regard between the White-Isaacs and the Barkers. Documentary data for the regard are scant in the early years, but in 1736 at the age of 14 years Thomas Barker was to record what appear to be the earliest known observations made by White. They occur in a manuscript diary kept by Barker[1] and read:

> 1736. — March 31.   A flock of wild Geese flew N. — G.W.
>   "   April 6.   The cuckow heard. — G.W.

Evidently friendship between the younger members of the family was already well established, and it developed in a number of directions. White himself visited Rutland again in 1742; at Easter 1746 members of the Isaac family were at Selborne; later the same year there was a further opportunity to renew acquaintance with the Barkers, since White spent part of June in Lincolnshire; while in July 1747 he paid the expenses for sister Anne's return journey from Stamford to Selborne via Oxford.[23] Moreover, since at least 1745, White's brother Benjamin had been learning the trade of publishing and selling books from a London relative of Barker, John Whiston; this was to lead to partnership with Whiston, and later the establishment of his own business specialising in natural history.[3]

At this time (1751), however, of far greater significance for White's development as an observer of the natural scene than the fact that his brother would publish the as yet un-dreamed-of *Selborne* were Thomas Barker's astronomical and meteorological interests. Principally, these centred on the weather at Lyndon, Rutland, for which he kept meticulous records over a period—for one man a uniquely long period—of nearly seventy years. The compilation of these data led to a knowledge of the English climate unmatched by any of his contemporaries, and enabled him to make perceptive comments about climatic changes during almost the whole of the century. In pursuit of this knowledge, Barker was the intellectual heir of his grandfather, William Whiston, who had succeeded Newton as Lucasian Professor of Mathematics at Cambridge in 1702.

The guidance White received from Barker concerning meteorological matters was therefore considerable and authoritative; and although the full significance of the guidance was not apparent until much later in the 1770s and 1780s, when White was seriously engaged in contributing to an understanding of the natural calendar, its influence was pervasive throughout his life. Such influence as Barker had did not, however, derive only from scientific matters: his personality and manner of life were also important. At the time of his marriage he had developed a reputation for 'extreme Abstractedness & Speculativeness', characteristics which might be expected to impair (but in actuality did not) a rapport established with White during his youth and sustained into old age. As a man in his sixties, he rode the 118 miles from Lyndon to Selborne in three days, White commenting to relatives that, although 'at every dining place, while the horses were baiting, [he] walked four or five miles in the heat in his boots with his wig in his hand', he arrived 'without the least complaint or fatigue'. This was in 1783, and the two friends had evidently not met for many years for White continues: 'He has still a streight belly, and is as agile as ever; and starts up as soon as he has dined, and marches all round Hartley-park. This morning [he] ran round Baker's hill in one minute and a quarter'.[4]

Consideration of the specific ways in which White profited in his meteorological studies from Barker is given later. At this stage it is sufficient to note that, although the first entry in White's *Kalendar* occurred the day after Barker's marriage, it would be wrong to read this juxtaposition as having great significance. Seemingly, Barker was at Selborne from mid-October 1750 to the end of February the following year, so there would have been many opportunities for him to encourage White to begin a regular record of activities in the garden; and further, the kind of entries White was to make in the *Kalendar* (in contrast to those of the later *Journal*) were very much subject to time and, more obviously, to the vagaries of the weather—a record of what is accomplished in a garden being decidedly subject to opportunity.[5] The most likely course of events therefore is that the initiation of the *Kalendar* arose as a result of Barker discussing in November and December 1750 his records of weather at Lyndon, White deciding, with the approach of a new season, to commence his own record at the first opportunity in January.[6]

Tacit approval for this venture would, White knew, also be given by the Reverend Stephen Hales, rector at Faringdon, a parish adjacent to Selborne, from 1722 to his death in 1761. Formerly, most commentators have attributed to Hales White's commitment to record-keeping, often basing their case on evidence provided by White's correspondence (subsequent to the publication of *Selborne*) with a sometime friend of Hales, Robert Marsham. Yet, although White is warm in his praise for Hales, and makes much of the familial nature of the friendship, he makes no acknowledgement (as Marsham does) of its influence as regards the keeping of records. Instead, he stresses almost exclusively the 'benevolent and useful pursuits' that Hales engaged in, and cites at length a series of anecdotes that demonstrates not only the man's humility but also the fruit of a mind 'replete with experiment'.[7] The proper place, then, for Hales in White's biography is not as prompter of the *Kalendar*, but as an exemplary figure concerned to promote human well-being and to solve some of the practical difficulties that beset daily living.

With the *Kalendar* launched, White began to think of his own future. Life as an occasional curate had discovered some disagreeable irreverence,[8] and, in spite of hearing that an appeal by his brother John against expulsion from Corpus Christi College, Oxford, had failed,[9] his thoughts began to turn again to the university. For the year 1752/3 it was the responsibility of his own college, Oriel, to nominate one of the two proctors—officers responsible for the discipline of junior members of the university. Although a non-resident member of his college, he was entitled to be a candidate and despite some hesitation he put his name forward. Possibly he hankered after the life of a scholar-poet (he was still writing poetry, and was to continue with the occasional translation from the classics), a more regular income, or a greater variety of activity and acquaintance than that found in the country; or perhaps he had an eye to secure preferment. Another thought may have played a part: that he might deal with the mischievous and high-spirited more effectively than had those in authority over his brother. But, whatever the reasons, the office was sought with Mulso's hearty support: 'I think of ye Office wth great Respect, & as I make no Doubt that You will execute it wth great Applause, I think it may be a serviceable Honour to you. I fling in my Vote to your accepting it.'[10]

With the prospect of duties ahead and of necessary residence at

Oxford, White spent part of the summer of 1751 making a tour into the north and visiting his newly-wed sister at Lyndon. The enthusiasm he showed for travel, customarily on horseback since he suffered from coach-sickness, invariably brought paeans of admiration from Mulso, who was something of a hypochondriac.[11]

Soon after his return White accepted new clerical responsibilities, this time as curate-in-charge at Selborne since the vicar, a Dr Bristow, was ill. During this period, he resumed work on the garden at Wakes, and at the end of the year concluded entries in the *Kalendar* with a summary of the weather for the twelve months. With the new year the demands of the proctorship were soon upon him: in March 1752 he resigned his curacy, departing from Selborne's 'green retreats' for the spires of Oxford, and thus initiated a period in his life best characterised (by adapting his own phrase) as one of 'Hectic Heat'. As proctor from April onwards he 'fully entered into the social life of the University, going to concerts on "choral nights," entertaining his friends with "A bowl of rum Punch from Horsman's," occasionally playing cards in the common room,[12] while entries [in his accounts] for cleaning his gun and the purchase "from Woodstock turnpike" of "a spaniel of the Blenheim breed" show that he still continued to shoot'. For part of this time he also held office as Dean of Oriel. True, at Christmas 1752 he was able to make a visit to Selborne, and to report to Mulso on his brother John's progress in constructing the famed zigzag path (which, much repaired, can still be climbed today) straight up Arbour Hill from the village through the hanging beech to Selborne Common above. A commitment to Oxford, however, is illustrated strikingly in the *Kalendar*: absence from Selborne inhibited collection of the necessary data, and the entries for the year (1752) conclude with no summary of the weather.

Because duties as proctor ended in May the following year (White went out of office on 2 May), it might be thought that detailed entries in the *Kalendar* would then be resumed; but White had other plans, and the year provides only the briefest of records. Mulso's suggestion (that White would so welcome being 'freed from the Confinement & Form' of office that he would 'exult & wanton' as a spring horse)[13] is colourful and, in many ways, accurate. After attending to affairs at Oxford, June was spent travelling: to Sunbury (to stay with Mulso), to London (to visit friends and relatives), back home to Selborne and on again, to Oxford; then, after a brief respite, he was off to Bristol, for a new experience—'a 5 weeks season at the Hot well'.[14] By female report, Mulso had understood Bristol to be 'very slightly commended' since there was no 'Room for Routs & Assemblies', but White seems to have sent a favourable account of his *crisping* (by the waters), and sufficiently enjoyed the occasion to return two years later for a longer season.[15]

By early September, however, the prospect of a settled curacy begins to emerge. Most probably, the arrangements had been made previously, since White's accounts[16] show that he paid for board at Durley from 8 September 1753. This was (and is) a small hamlet no more than a convenient ride from Selborne and under the jurisdiction of James Gibson, rector at Bishop's Waltham, with whom he lodged.[17] The responsibilities of a country parson were considerably less onerous than those of proctor and, as can be inferred again from White's accounts for the period, he put his freedom to good purpose. The change of focus from city to country, from proctorial authority to village curacy, is sharply indicated by the way he spent his money, which becomes predictive of the evolving pattern of his life. In mid-October he resigned his deanship at Oriel, and on 21 October he paid bookseller brother Benjamin for a new edition of Philip Miller's famous *Gardeners Dictionary*.[18] In fact, the volume was in exchange for an earlier edition obtained in 1747. Together with the exchange, he bought a copy of John Ray's *Methodus*,[19] and both purchases indicate not merely a renewal of interest in the natural world, but something much more important: a decision to turn his back on the prospect of academic and clerical preferment, and instead 'to pay a more ready attention to the wonders of the Creation, too frequently overlooked as common occurrences'.[20]

The threads of gossamer observed in 1739 were now, one might say, beginning to work their mystery.

CHAPTER FOUR

The Garden

The main source for discussion of White's garden is *The Garden-Kalendar*. At first sight, it is a selective, repetitive, and noticeably erratic record of gardening activities at Wakes over the 16-year period of its compilation.[1] As such, it would seem of small importance: many gardeners have kept similar records, and, apart from a certain antiquarian interest, White's records it may be argued have little to commend them above those prepared by others.[2] To determine so, however, would be to miss the very reasons that make the document significant for the serious student of White and for an understanding of *Selborne*. In studying this last, most commentators emphasise the long gestation of the volume (almost twenty years from the time the idea was first suggested), and a few point to the illumination that can be gained from reading alongside *Selborne* White's *Naturalist's Journal*, which he kept from 1768 until a few days before his death in June 1793.

These are perfectly reasonable ways of extending a reading of White, and cannot be ignored. Yet, if we are to grasp how he came to begin the correspondence with Pennant and Barrington (in 1767 and 1769 respectively) that was to result in *Selborne* itself, close attention must be given to White's knowledge and training at the time the correspondence began. Such a task demands that we analyse records assembled before, say, 1768.[3] In accomplishing *that*, we shall establish the developmental importance of White's pre-1768 record-keeping, and be able to follow the process by which the country curate of the 1750s with (in Mulso's evocative phrase) 'dirty Boots & dripping Bob'[4] was to become the widely celebrated author of the book that for the past 100 years and more has grown into a representative symbol of the English delight in a varied, temperate, and enriching, pastoral feast.

Before we begin to explore the actual records, it is helpful to understand the kind of garden White worked in and enjoyed. Guidance can be gleaned from many of his letters, and from the accounts, as well as from the letters of Mulso. This last might seem a somewhat restricted

23

source (Mulso visited Wakes, in the early years of the acquaintance, at almost ten-year intervals), but Mulso's views enable us to see how White's endeavours were regarded by a contemporary. Although not a great traveller, and certainly no gardener, Mulso shows that, along with most educated people of his time, gardens and their appreciation were absorbed into the culture of the age. In part this is because much of the thought of the century can be seen as a movement from the public to the private, from the autocracy of a monarch to the democracy of the people, from servility to independence. Such themes, and many others, are vividly illustrated in the changing theory and practice of garden design. To Mulso's great credit he will help our mind's eye to see the garden at Wakes in a way better approaching his friend's intentions, and in so doing we shall more easily relate White's accomplishment to the practices of his times.

A convenient place to begin is with the ha-ha. Some thirty or so yards from White's great parlour one can still stroll along his 'terrass', peer into the ditch, prod the massy stones of the wall, and cast one's eyes across 'the great mead' to the beech-clad hanger beyond. The detail of its construction is given in the *Kalendar*. Digging the ditch and building the bank took 16 days and was completed on 31 March 1759; two years' settlement was then allowed, and the building of the wall was started by 'Long the mason' and completed on 24 January 1761, White noting that, because of the huge stones ('blue rags') that had been used, there was 'double the Quantity of stone usual in such walls'. The height of the wall was, in the main, 4 ft 6 in, and the cost of construction 'exclusive of carting the stones' was £1-8s.-10d. Over the following months the ditch was properly 'sloped' and the terrace levelled and turfed; and 'a finishing hand' (trimming the edges of the turf around the terrace, and mending some patches that had not taken) completed the task on 11 April.[5]

Whereas the detail of this account is unexceptional, it is worth noting that all the distinctive stages of construction are carefully recorded. One of the frequent characteristics of observations that, in a scientific sense, are of value is the observer's ability to break down a certain process into distinctive stages, and to record them accurately. To recognise and name these stages is an important attribute for many scientists, and it is interesting to see White demonstrating this ability on a structure as immobile as a ha-ha. It may, perhaps, be objected that to record in the *Kalendar* acts as necessary as digging, building, and levelling is simply to state the obvious—and there is truth in that. The point, however, lies (at this stage in White's development) not so much in the originality of what is observed, but in the willingness to accomplish the recording with a proper care and attention to accuracy. Without this trained capacity much original work in science would never be accomplished, and White was to achieve what he did because of his willingness to record. Several examples of these procedures will be examined later, especially in connection with the meteorological and other data given in the *Naturalist's Journal* (1768–93), but other examples occur in the *Kalendar* concerning garden operations and they will be noted shortly.

One further feature of White's account of the construction of the ha-ha — the comparative comment about the quantity of stone used — is also of interest. The judgement is evidently derived from the mason, who, although a local man, seems to have possessed knowledge enough, either from experience or from talk with fellow masons, to express the opinion he does: in a non-metropolitan society such as east Hampshire, the building of ha-has was by 1760 not so unique an 'improvement' as it might seem. Confirmation of this view is found, if negatively, in White's own writings: at no time in the *Kalendar* do we find any explanation for the 'improvement'—and this for a construction that a modern garden historian rates as 'a technological advance in the craft of gardening which is quite exceptional', to be compared solely with the invention of the fern-case and of the lawn-mower.[6] That White finds any rationale unnecessary is, of course, in part a factor of the kind of record he is making, but it can also be taken as an index of his period. By the time of the building of the ha-ha at Wakes such devices were a common feature of mid-century gardens, representative of an attitude to the relationship between the garden and the wild advocated by philosphers and poets earlier (Shaftesbury, Addison, Pope), and already put into effect by garden-designers such as Bridgeman and Kent.[7] The relationship now pursued was wholly at variance with approaches to gardening in the preceding century, at which time a garden had been seen as a formal, enclosed construction, designed absolutely by man in the face of the uncompromising encroachment of which wild nature was capable. Such an attitude was expressed in gravelled walks, in knot-gardens, in topiary: it has been termed 'architecture on the flat', and was a style displayed in many of the college gardens at Oxford with which White was familiar.[8]

With the introduction into garden design of the ha-ha (and White will have first seen one at Blenheim and Stowe during his Oxford years),[9] the landscape garden was born. This birth extended in two ways the notion of what a garden might be. Firstly, the sunken boundary, a ditch instead of a hedge, enabled what was outside the garden to be viewed with as much appreciation as had formerly been given to what was inside the garden; secondly, concomitant with the possibility of now being able to see what was beyond the boundary, came an invitation to accomplish just that—not solely in prospect (by looking), but by venturing forth physically with one's feet. As the century advanced, there were many variations on this theme of extension. A particularly simple (if costly) form pursued in the second half of the century by many great landowners was to employ Lancelot 'Capability' Brown to impose an understanding of what was beautiful on an entire landscape. Perhaps 'impose' is a strong term, for Brown took care both to work with nature and to ensure that final effects appeared natural. Nevertheless, deliberate imitations of natural scenes were achieved by drastic means: anything merely useful, such as a kitchen or fruit garden, was banished out of sight to the fringes of the estate; grass was brought to the very footings of the mansion; where there existed a valley a lake was created (Virginia Water in the 1750s being a good extant example); and on hills—usually adaptations of existing hills—were planted belts of trees with contrasting appeal. The resources required to make this kind of scene were immense.[10] Compared with the 400 acres of Stowe, there were a mere ten at Wakes: and the Whites had a living to make and a household to keep, all within restricted means. The solution at Selborne had to be something different.

What that solution became (for, as mentioned in an earlier chapter, the 'home' garden had been laid out originally by White's father) is best seen as the union of necessity with fancy, of the practical with the ornamental. In a sense, White's approach to these two contradictory demands was similar: to both he devoted attention over a recorded period of forty years, and in that time not only extended the physical boundaries of the garden but also enhanced its imaginative rewards. And yet, scrutiny of *Kalandar* and *Journal* shows that, whereas the routine practical care of the garden proceeded with undisturbed stability over the forty years, the decorative and fanciful element was subject to continuous development. Methods of cultivation available to a village clergyman were of course much the same at the end of the century as at the beginning, controlled largely by the vagaries of the weather, by the onset of pest and disease, and by the then scant knowledge of the physiological demands of many species and varieties.[11] In such circumstances any 'improvements' that occurred were essentially gradual and hard-won. By contrast, some aspects of garden design could be accomplished far more easily, and it is in this area that we can watch White making 'improvements' and refinements throughout his gardening lifetime, from 1751 to at least 1791. The chronological detail of this activity is given in outline later (see Appendix C); for our purposes we can more appropriately address the major features of the garden, and then interpret them in the light of contemporary, public theory and practice.

The union of necessity and fancy is nowhere better illustrated than in the ha-ha, where the utilitarian value of ditch and wall becomes aesthetically enabling — in that the observer's sense of space is significantly extended, what was beyond the boundary seeming to become an integral part of the garden. In fact, for White, there was no 'seeming' to be suggested, for a striking feature of his garden design was his success in colonisation of the surrounding terrain, certainly of the village, and often even further afield.[12] The expressive source for this colonisation is so rich that we shall have to consult White as poet as well as gardener, and as classical scholar as well as clergyman; but the best place to begin is with the natural scene itself.

The topography of the village is in the main unchanged over the past two centuries or so, and although there are differences (and very striking ones as soon as one looks at more than the superficial) the reader of White's description of the main features (*Selborne*, Pennant 1–9) readily acknowledges a high level of continuity.[13] In essence two features predominate: the extensive tree cover, and the undulating ground. Both of these were commented on by Mulso. As early as 1746 he directs letters to White as someone who lived 'amongst ye Hills', and later gives us a glimpse of his correspondent's own inclination. In an account of a day's 'Jaunt of Pleasure' in the environs of Windsor Great Park, his judgement is that the scenes would afford White 'a Subject of Entertainment', and continues: 'I know it will please You the more, because you will own the Ground to be irregular'.[14] Evidently, White took little pleasure in the regular and flat. As for trees, certain apparently commemorative plantings by White have been mentioned earlier; and Mulso's evidence helps us to extend further our understanding of this interest. Mulso had been given news by White of the planned despoliation (probably routine felling) of the woods on a hill adjacent to the village, and responded:

> I grow outrageous at the Injury or Sacriledge which is going to be committed on Noar Hill; I cannot bear an Intention of profaning these venerable Shades; and when I feel ye chilling awe which You prevent me in describing, I reflect on ye wisdom of the Artifice of ye old Idolatory, who placed their Altars & Images in such Scenes, where the Place created the Reverence which the grotesque Figures [the Druids] could hardly have inspir'd . . . [and where the] striking Gloom seem'd to justify ye Superstition which call'd it Holy.[15]

And there, of course, is the clue to the kind of extended garden White was to create. The natural scene at Selborne, with its 'venerable

Shades' and 'striking Gloom', was indeed a place of 'chilling awe' and 'Reverence'—not, it must be understood, because of the 'old Idolatory', but because of its capacity to 'enlarge' the observer.[16] Such a notion, of a view 'enlarging' man, would, one might think, apply most aptly to the grandeurs of the natural world; but for White, although he responded to 'that chain of majestic mountains [the Sussex downs] with fresh admiration year by year', he did so in terms that values 'the shapely figured aspect of chalk-hills in preference to those of stone' which latter he saw as 'rugged, broken, abrupt, and shapeless'. His elaboration of this preference is worth noting: on each occasion he saw the downs, he wrote that he perceived

> somewhat analogous to growth in their gentle swellings and smooth fungus-like protuberances, their fluted sides, and regular hollows and slopes, that carry at once the air of vegetative dilation and expansion.[17]

That White's mind should have this particular cast is what the reader of Mulso, or indeed of White himself, might expect. At frequent intervals in his letters Mulso points to White's love of trees, one of the most striking judgements occurring at about the time White commenced his proctorship. In congratulating his friend, Mulso opines that he has 'paid the University a great Compliment' in being willing to carry the office of proctor since 'the green Retreats . . . of Selbourne afford more serious Pleasure to your contemplative Mind'[18] than do the public responsibilities of office. Comments of this order suggest strongly that the garden White sought was of a special kind, a place apart from the hurly-burly (so beloved by Mulso), where he could quietly contemplate the mysteries of life. Such places are often termed sacred groves, but although there were trees in abundance at Selborne there would seem to be no structures, such as a shrine or perhaps an altar dedicated to an appropriate deity, to justify calling the groves sacred: certainly, White was to erect nothing of the sort. That the green retreats of his native village were not to be enhanced by a shrine or altar is at first surprising, but we need only a moment's contemplation to deduce what must be the reason: both shrine and altar were already in existence. In fact, each was of such antiquity that, when White came to record the ecclesiastical history of the parish (in the second part of *Selborne*, known as 'The Antiquities of Selborne'), the necessary research took several years: I refer, of course, to the several pre-Reformation ruins in the parish, and to Selborne Church itself.[19]

Although the existence of shrines and altar substantiate in a strict sense the sacred nature of the Selborne topography, it is, clearly, the natural beauties to which White responded most joyously. This is made plain in his longest poetic production, 'The Invitation to Selborne'.[20] Here, after a description of the village's 'boldest beauties', he concludes with a paean of praise to the 'pendent forests, and the mountain-greens' where 'nature hangs her slopy woods' through which the 'Rills purl between and dart a quivering light': trees, rocks, and streams—a sacred grove indeed.

Sanction for this kind of approach to a garden is found in classical and biblical literature. For the latter, the paradisal, natural garden of Adam and Eve is the archetype. Somewhat strangely, this is not mentioned obviously by either White or Mulso, both of whom prefer frequent reference to seeing Selborne as analogous to some of the famed sequestered gardens as recorded in Greek and Latin literature. Indeed, Mulso goes well beyond analogy, and makes metaphorical identifications. As early as 1750, for example, he writes of how White seeks to 'creep into the Shades, or shelter in the *Nidus Acherontiae* [the nest of Acherontia]; by which Name I think you have christen'd your Arbour in ye Hill', and in 1767, when pressing his friend to come to Witney, he urges: 'Leave . . . your Tiburni lucum [grove of Tiburnus], & deign to visit Us.'[21] White himself is usually much more restrained about the delights of Selborne and, especially in later years, uses classical quotation in illustration of a continuing process; that is, he shows the applicability of many of the observations made by the poets of the past. An especially notable example of this occurs in the *Naturalist's Journal* in 1788. On four occasions in that year lines from Virgil's *Georgics* are cited: in two instances White comments that an observed event has put him 'in mind' of the quoted verses, but in the others he records that something he had witnessed 'verified' an observation of Virgil's, the most famous of these last being the note of 4 October:

> The prodigious crop of apples this year verified in some measure
> the words of Virgil made use of in the description of the Corycian
> garden – – –
>> "Quotq[ue] in flore novo pomis se fertilis arbos
>> Induerat, totidem autumno matura tenebat
>> [And whenever the tree which is fertile with apples
>> had dressed itself in new flower, then in autumn it
>> retained the ripe fruit]."[22]

But this is to anticipate, for the Corycian garden was a practical, cultivated garden and, before we consider similar dimensions in White's garden, comment must be made on what I termed earlier the union of the practical with the fanciful.

At bottom, what White loved was the natural beauty of Nature, in which the beneficence and ingenuity of God's wisdom for man could be endlessly sought. To this vision, identified in White's mind with Selborne itself, man added two dimensions: on the one hand the day-to-day necessities of husbandry and gardening, on the other the modification and interpenetration of the natural scene in ways that enabled man to obtain even greater delight from Nature than if he stayed in his drawing-room. In a measure, this notion of interpenetration was a matter of providing the onlooker with a variety of scenes, ideally as a series of unexpected pleasures. The desire to offer this level of variety, mimetic of the natural scene itself, had been well expressed by Pope, who advocated:

> In all, let Nature never be forgot.
> But treat the Goddess like a modest fair,
> Nor over-dress, nor leave her wholly bare;
> Let not each beauty ev'ry where by spy'd,
> Where half the skill is decently to hide.
> He gains all points, who pleasingly confounds,
> Surprizes, varies, and conceals the Bounds.[23]

Scenes such as Pope describes were not a device of unbounded Nature on her own account—at least not entirely so. What Nature provided could also be 'methodized', and her beauties sought. To accomplish that kind of variety demanded that the natural scene be peopled: that is, that men and women should actually live in and frequent her leafy groves (or at least seem to do so)—and White saw to that too.

How, then, were such ideas articulated on the ground? Firstly, by controlled vistas: the advantages of the ha-ha have been discussed earlier, but White also constructed a vista 'of six Gates one above another in Perspective', something which Mulso supposed to be an original idea, terming it 'as new, as it is agreable'. Indeed, so new and strange did he find the idea—as a planned, constructed vista—that he discussed it with friends and reported subsequently:

> we have not reformed our Ideas of your six Gates, which we conceive to be pretty but to sound oddly in Description: We take them to belong to Fields which thro' an Opening are seen in Perspective, One above Another, yet not so as to join; & the Image itself is not ridiculous in our Minds, but new to our Observations.[24]

A rural clergyman had something as early as 1754 to teach metropolitan Mulso it would seem! Yet these vistas, even with the addition of brick walks and benches, vases and an obelisk, needed complementing, for they were on level ground.[25] So, secondly, White created structures that elevated the onlooker, and widened the prospect: these were a conical mount, a turfed mound around his great oak, a zigzag path so as to ascend the hanger (although the actual construction of this was directed by his brother John), and later a less strenuous ascent through the hanger known as the Bostal. Each of these devices gained height for the observer of the Selborne scene, and thereby enhanced appreciation of its richness and variety. But the perambulating admirer would also require rest, relaxation, and refreshment, and so, thirdly, White erected high in the hanger a hermitage—in fact, two.[26] One of them, adjacent to the top of the Zigzag was in position quite early in White's 'landscaping' of the village; the other was built much later, in 1776, and was visible from Wakes garden.

The notion of a hermitage had several origins, both classical and mediaeval, all transmitted by aspects of high literature. Whilst the core of the creation was the appeal of the meditative life (Milton's appeal in 'Il Penseroso':

> Find out the peacefull hermitage,
> The Hairy Gown and Mossy Cell,
> Where I may sit and rightly spell,
> Of every Star that Heav'n doth shew,
> And every Herb that sips the dew . . .[27]

being a typical expression), White sought two other functions, both of them public. In the one function, we must see him as an engineer of illusion, rivalling not perhaps the mysteries of a Prospero, but certainly taxing the credulity of some of his younger relatives and friends. In pursuit of the illusion White used to arrange that the Hermitage was occupied, for the sake of visitors, by his brother Henry (see Plate 2). So successful was this device that at one time Mulso was to write:

> The Secret is out with regard to *the Old Man of ye Hill* . . . But Jenny [Mulso's daughter] is highly dissappointed; She had amused her Thoughts with this uncommon Expectation; & tho' her Curiosity longed to be satisfied, yet She enjoyed more from her Curiosity than She has done from the Solution of the Affair. I was sorry that my Wife judged it best to put a Stop to her Inquisitiveness [Mulso's wife had told Jenny the truth of the hermit], for it was innocent & often ingenious; But her speaking of this to Others, & our being obliged to disguise the Truth before her, gave Us the air of Romancers.[28]

Allied to this device of solitary contemplation in simple surroundings was a quite different convention, that of pastoral pleasure pursued by shepherds and shepherdesses. For White, this convention reached its climax in 1763, a note in the *Kalendar* for 28 July that year reading:

> Drank tea 20 of us at the Hermitage: the Miss Batties [three young ladies from London, at the time summer guests in Selborne of Mr Etty,

at the Vicarage], & the Mulso family contributed much to our pleasure by their singing & being dressed as shepherds, & shepherdesses. It was a most elegant evening; & all parties appear'd highly satisfy'd. The Hermit appear'd to great advantage.

The success of the occasion, indeed, of the whole vacation, is confirmed by a somewhat breathless entry in the diary of one of the impressionable young ladies. 'Adieu happy Vale enchanting Hermitage', she wrote; 'here the scene closes — the play is done — the pleasing dream is oer & tomorrow I must awake & find myself in London'.[29] Although these activities at the Hermitage were so appreciated, a more permanent use was as a location for a simple repast. Of this second function, White again is planner and organiser: though not as formerly noted within a literary convention, but as part of a secure custom of a host feeding discriminating palates rather than the literary illusions of young and impressionable guests. I refer to the special gastronomic pleasures provided from his own resources, namely to White's passion for melon-growing, and sharing his success with friends. The documentary data for the Hermitage being used in this way, as the venue for a melon banquet, are scattered through White's writings. One of the most informative accounts occurs as early as 1758. In the *Kalendar* that year for 12 September is the note:

> Held a Cantaleupe-feast at ye Hermitage: cut-up a brace & an half of fruit among 14 people. Weather very fine ever since the ninth.[30]

But pleasures such as these had to be won, in many cases in the face of severe conditions for successful gardening; and the spirit of the age, and White's ingenuity, found an answer for that also, which was to demonstrate his acknowledgement that the historic continuities were not exhausted by an appeal to civilised simplicities and pastoral play.

Quite where this fourth element of White's garden design is derived from is not recorded, although what he constructed was so fashionable that only a year or so before periodical literature had carried a fierce criticism of what he intended: the erection of a statue of Heracles.[31] That he chose this particular deity from the pantheon of heathen possibilities, and that he wished to display any such figure, needs some explanation. White's version of the Heracles legend relates in particular to the eleventh labour: the winning of the Apples of the Hesperides. The Garden where the apples grew belonged to Atlas, and the tree of golden apples was guarded by Ladon the dragon. Because of the popularity of the legend and its appositeness, many gardens contained statues of Heracles, one famous Italian garden at the time of the Renaissance deploying Heracles statuary throughout the garden. The point made at each garden which showed a Heracles was, of course, a claim that the garden was a Hesperidean Paradise *re-found*.[32]

How far White subscribed to this kind of interpretation of his statue is not easy to establish; and we should bear in mind that east Hampshire, and Wakes garden in particular, were not very public places. Also, White had very few visitors—outside the family that is. A more plausible reading of what he was about is something much more personal. The erection of the figure afforded him a quiet satisfaction: as a private symbol of the enduring labours to be accomplished in the garden if man is to win fruit from its taxing demands. Whatever the reading, however, we know that the statue (no more than a wooden cut-out) was being worked upon during October 1757, and that it was set up in the following January:

> Set-up about 20 Yards into the Hanger, in a line with the six Gates, a figure of the Hesperian Hercules, painted on board, eight feet high, on a pedestal of four feet & an half. It looks like a statue, & shows well all over our out-let.[33]

No doubt the simplicity of the structure was in part an expression of the financial status of the Whites, but it also conveys a more rigorous attitude: money for essentials (such as enlarging the garden, or paying for an indenture),[34] but not for frills. This union of the simple and the pleasurable, bound together by virtue, is expressed directly by a passage in Milton's *Comus* that White assuredly knew:

> . . . musing meditation most affects
> The Pensive secrecy of desert cell,
> Far from the cheerfull haunt of men, and herds,
> And sits as safe as in a Senat house,
> For who would rob a Hermit of his Weeds,
> His few Books, or his Beads, or Maple Dish,
> Or do his gray hairs any violence?
> But beauty like the fair Hesperian Tree
> Laden with blooming gold, had need the guard
> Of dragon watch with uninchanted eye,
> To save her blossoms, and defend her fruit
> From the rash hand of bold Incontinence.[35]

What is under threat in the passage is actually chastity: whereas no one will threaten the simple hermit, a paradisal simple way of life can be threatened and needs protecting. Hence, as well as the Heracles, White had his dragon, although its form is not known.

With these figures, one might expect White's representation of the classical to be complete, but that is not so. As well as Heracles and the dragon, there appears to have been some structure termed 'the Cynic

Tub'. This last phrase is frequently used in literature to refer to the Greek philosophical school known as the Cynics, whose name derived from the behaviour of their most famous representative of lecturing from a tub. As a group the Cynics were contemptuous of wealth, ease, and luxury, regarded virtue as the prime basis for happiness, and promoted the practical virtues. White's wish to have in his extended garden some such figure accords wholly with what we know of his morality and of his commitment to Selborne and its simple pleasures.

That it is right to stress White's love of his village for the beauty of its prospects, and the simplicities of its daily routines, is strikingly and visually shown in *Selborne* itself. As originally published, the volume has a frontispiece that relates to the totality of the book and a separate vignette for each of the parts. In many modern editions all these are often omitted, with a significant loss of data for understanding White's attitude. The frontispiece is an engraving of a view over Selborne from the northeast (see Plate 1). The direction of the view is important, for it enables us to share the whole of the village; the vantage point is elevated; on the hanger may be discerned the Zigzag, and the hermitages; and—this surely a touch of the exotic—in the foreground stands part of a 'Turkish' tent used for summer entertainments.[36] In the engraving, the posed figures are shown sharing with us (or directing us perhaps) to a view of the whole that says with an alert satisfaction: this is my garden; I have fashioned it thus; dwell quietly here and be content.

The vignette for the first part of the volume, the 'Natural History', extends this theme of contentment. The scene shown is of the hermit at his hermitage, with a prospect of the valley beyond. The legend is simply descriptive: 'where the Hermit hangs his straw-clad cell'. The significance lies in the mottoes, one in Greek, the other in Latin. The former reads in translation: 'It is a rough country, but a good place for rearing young men. I for my part cannot envisage anything sweeter than one's own country'; and the latter: 'In short the whole of that territory of ours is rough and mountainous, [but] dependable, trustworthy and a protector of her own people.' More certain adherence to the values enshrined in Selborne would be difficult to state. Yet both are tinged with sadness. There is a fragment of verse White wrote:

> Sweet is the splendour of the morning sun;
> And sweet to see the gently heaving main;
> Sweet is the vernal face of hill or dale;
> And sweet th'effect of fertilizing stream:
> But far more cheering, far more lovely scene,
> Is to the childless man, wrung with desire,
> The sight of new-born children in his house.[37]

Despite the perfection of the territory White knew so well, he was never to share it with life of his own loins, for he lived and died a bachelor. From this sadness, he sought solace in nature's beauty, and in his own garden: and it is to the plants he grew, the fruit he picked, that attention must now turn.

CHAPTER FIVE

The Garden-Kalendar

Scanning the *Kalendar* for the first time, especially for the eight years 1757–64,[1] it is easy to become both bemused by the variety of endeavour and absorbed by the power of the prose. Every sentence, almost every phrase, rings with an energy and confidence that extends far beyond what is usually associated with garden records. Consider, for example, these entries that were made for the first days of April 1761:

April 1.   The Succade-bed continuing too hot, I ordered a pole to be thrust quite thro' the bed under each hill, so that one might see thro'. One hill being more furious than the rest I had the plants (top of hill & all) taken-off in a shovel, & the hill new-made-up with cold earth. The plants grow, & are not yet injured.
Grafted-in two cuttings of M: Middleton's espalier-Crasane, instead of the Ringmer ones, which were canker'd, & bad: left one Ringmer one.
Planted 6 basons of double larkspurs in the new-garden borders.

2.   Sowed in the seed-bed in the melon-ground Battersea Cabbage-seed, Savoys, borecole, leeks, Holyoaks, stocks, carnations, & sweet Williams.
Bright sunshine, with an east-wind, & very high barometer.
The Ground is bound like a stone.
Hoed, pronged, & cleans'd all the home garden, & borders, during this parching season.

3.   Planted a Garden Lathyrus on the new bank between every two asters, & p: sunflowers.
Stopped-up the vent-holes in front of the Succade-bed; left them open behind. Most of the plants look well.
Sowed a great Quantity of Cucumber-seeds for the neighbours.

4.   Sowed tree-prime-roses, beet, & some seeds of a red Cowslip in a pot.
Sunny, burning weather.
Dress'd artichokes: in that hot weather the beginning of December they sprouted thro' their ridges, & continued growing very much the winter thro'; & have now vast greens.
Hoed & cleansed the grubb'd ground in the meadow.[2]

Here, briefly and simply, is the quintessence of White's style that was to serve him so well in the future when he communicated with others about natural history. Although the extract makes no mention of events that are instantly memorable (the building of the ha-ha and fruit-wall, the melon feast at the Hermitage, observation of the 'seven swallows, the first this Year, playing about James Knight's House' . . .),[3] a relish for the sheer worth of the activities described shines through every line. In part, this is because of the subject matter: man's dealings with the fruits of the earth. It is also because of White's particular mode of recording (many of the entries are introduced by reference to an accomplished act — grafted, planted, sowed, hoed, pronged, stopped, and so on). Both these factors suggest a high level of productive activity; but there are other reasons for our sense of bewilderment.

The most noticeable reason is the decisiveness of the record. Admittedly, the account deals with facts, and facts can scarcely exude an air of indecision because they derive from human decisions; but occasionally, at least, one might expect the record to report some prior debate about the best way to proceed, about the most urgent of a number of tasks, or perhaps about whether conditions were appropriate for this job and that. To write such a journal would, of course, convey a relationship to the garden quite different from what White accomplishes, and yet one advantage would have accrued: as a human document we might, as gardener-readers, have found the *Kalendar* less humbling than many of us now do. The decisiveness, however, is supported by other factors, best described as narrative particularity and collaborative exuberance, both of which compel attention — as if Nature herself was urgent on her own behalf, inviting co-operation and skill in assisting in the success of the varied plantings at each and every site named in the record.

Such terms invite explanation. In the four-day entry given above, at least ten different sites are explicitly present; and scrutiny of the *Kalendar* entries for, say, February–March 1759 yields over two dozen locations at which considerable activity was accomplished. The sense of scale that these figures suggest is vastly multiplied if we think of plant species: some twenty are referred to in the April entries mentioned, and a count across the years would give several hundred. To grasp the significance of such figures is not easy, either numerically or in terms of variety, for the figures include material that can be classified in different ways. Comparison with a popular gardening manual of the time, *Every Man his own Gardener*,[4] shows that, although Wakes was buried deep in the Hampshire countryside, White was familiar with, and growing, a very high percentage of what was then readily available. True, he had no greenhouse[5] so the number of stove plants mentioned is very few, and the size of the garden meant prudent planting of what Abercrombie calls deciduous and evergreen shrubs and trees for 'furnishing Noblemen and Gentlemen's Gardens and Plantations'; but in each of the five other classes White scores highly. For example, of the fifty (excluding 'Sweet and Pot Herbs')[6] 'Kitchen-Garden or Esculent Plants', White grew nearly all; and of the 19 fruits in Abercrombie, White grew every one, including peach, nectarine, walnut and fig, medlar, mulberry and quince, filberts (of which he had a hedge) and sweet chestnuts (which Abercrombie names as the 'manured, or Spanish' sort). For many of these plants White grew several varieties (for example, 'a dozen types of bean and melon, six kinds of plum, and seventeen sorts of pear').[7] The record is notable for its good sense. Although enormous industry is expended over twenty or so years on melons, he seems eventually to have decided that the reward scarcely matched the cost or effort,[8] and he never experimented with pineapples—but was content to admire those in the hothouse of a friendly neighbour, Sir Simeon Stuart.[9]

What he did take care of was the necessary provisioning of his household, and in quantities that to many of us today, particularly at a first reading, must seem wasteful, even extravagant. The most astounding records relate to the number of cucumbers that were harvested. For instance, as late as 1789, White noted: 'Gathered a *bushel-basket* of well-grown cucumbers, 238 in number'—and this on a single day's cutting.[10] The necessary plantings of winter vegetables are equally arresting. In June 1758, the *Kalendar* records:

> Prick'd-out 600 Savoys, & 350 Boor-cole-plants . . .
> Planted-out all the leeks . . . about 200.
> Planted 100 Cabbage-plants . . .
> Pricked-out an 120 Roman-Broccoli . . .

and in August 1757:

> Planted-out . . . near a thousand savoys . . .

Yet successful planting is no guarantee of a satisfying outcome; and White knew that. On occasion the yield from his garden delighted him, but on other occasions he had to record: 'Gathered my apple, & pear-crop, which consisted literally of one Golden-pippin, & one Cadillac [pear]';[11] and the cabbages cited above were 'in the room of those planted May 31 . . . which were dead' (probably drowned). In some circumstances the prospect of failure was averted by diligence and foresight:

> The caterpillars have been pick'd off the savoys several times: those
> that have not used that precaution have lost every plant . . .

and sometimes by experiment:

> Those melon-plants that were once seized with a mouldiness constantly
> dy'd away by degrees, 'till they were quite devour'd by it; except those
> plants on which I tryed the experiment of clipping-off the infected part
> with a pair of scissors: when they recover'd & afterwards grew pretty
> well.

On other occasions damage occurred that it would have been difficult
to prevent. In autumn 1763 (24 October) White wrote: 'Hares or
some vermin have gnawed almost all the fine Pheasant-eyed pinks, &
the new-planted cabbages'; and in the following autumn even worse
took place:

> In the night between the 16: & 17: [September] my melons &
> Cucumbers were pulled all to pieces; & the horse-block, three hand
> glasses, & many other things were destroy'd by persons unknown.

What, perhaps, is most noticeable about some of these entries is the
precision of report—not always, of course, in numerical terms, but in
emotional. At no time do we feel that any sense of outrage, disbelief,
or even a momentary misery is permitted to cloud the accuracy of
report. Listen to these two entries, both from 1759: 'Continued
picking vast quantities of slugs from the french-beans, which are in a
poor way', and (an entry only a fortnight earlier, on 15 May): 'All my
Savoy-seed, & Boor-cole fails this Year: not one plant appears'. There
is no self-pity here; the eye is kept keenly on what is being observed. It
is a quality that will serve White well later, as a naturalist.

Satisfaction of the palate, through provision of basic necessities,
was supplemented by other plantings; for the eye also must be
pleased. Of Abercrombie's three lists of flowers, White grew most that
are recommended. Not surprisingly, the hardy perennial and biennial
'Flower Plants',[12] which Abercrombie distinguishes from those
'Plants as may be raised from Seed, and which merit Places in a
Garden as ornamental Plants', are usually mentioned only at times of
receipt, planting or other operation, or when they are in flower.
Lastly, there are the annuals. This is the group which Abercrombie
subdivides into three classes, the third of which comprises those
annuals that are hardy and hence suited, in most instances, to being
'sown in the Places where it is designed they shall flower'. The other
two classes require a hotbed, and it is noticeable that White grows
plants in both of them.[13] The bed is usually termed 'the annual bed',
and in 1761 for example it took '7 barrows of dung' and was to be

covered with 'the biggest one-light frame'; here, on 11 April (three days after preparation of the bed), was sown:

> dwarf sunflowers, Marvel of Peru, Ba[l]soms, China Aster purple, & white, Fr: & Afr: marrigolds, Pendulous Amaranths, & Convolvulus minor.[14]

Quite where all these were later transplanted we are not told, but on 17 June White arranged 'the winding-border over-against the fruit-wall with tall annuals behind, & a row of China-asters before', and in spite of their being 'sadly scorch'd by the heat' (25 June) he was able to note on 28 August, after having been away on a visit, that 'the annuals [were] very handsome & very strong'. The season, in this respect at least, was a success. And *that*, perhaps, hints at what earlier has been termed the narrative reward of the *Kalendar*. Time after time, in connection with place and plant, one can follow over a season, and in some instances over many seasons, White's endeavours and experiments—for his tastes and his skills change. Two examples must suffice: one of a plant that occurs early in the record but only in two years, and another of one that is grown from the mid-1750s until 1793.

Mention of amaranths[15] occurs at several points in the *Kalendar*, but the first reference is an entry in the hand of White's brother Thomas for 16 April 1752: 'Sowed in the New Garden on the Border by the Brick-Walk, Love lies aBleeding . . .'. How these plants, *Amaranthus caudata*, fared is not recorded: White was much away from Selborne from May 1752 until the autumn of 1753 and the record is incomplete. In the following year (1754), however, in addition to mention of love-lies-bleeding, which White names as 'pendulous Amaranths' and which he obtained as seedlings, considerable energy was devoted to growing cockscombs, *Celosia cristata*. The record, when assembled, is extremely full and begins in March:

12 March	Sowed . . . one pot of Cockscombs [in a very deep hotbed made on 5 March] . . .
19 March	Sowed . . . One pot more of Cockscombs.
[– March]	The first sown Cockscombs appear'd about the 21st. came-up very thick . . .
25 April	Planted-out about 20 of the best Cockscombs on the upper side of the Cucumber, & two-light melon-frame.
8 May	The Cockscombs wonderful forward, & stocky; & have showed bloom ever since the end of April.
21 May	Made a good strong hot-bed to finish-off the Cockscombs with: plunged 10 large pots in the bed, & half-fill'd them with fine earth.
22 May	Moved ten of the best Cockscombs into the large pots in the new beds: the plants were taken-up with a sheet of

	tin with a deal of earth, & well water'd. The plants very fine, & forward, & in good bloom; & 22 inches high. Two old frames placed one on the other: & the bed beginning to heat well.
5 June	The Cockscombs full 28 Inches high; the combs very broad, & the stems very stocky.
15 June	Cockscombs 3 feet high the tallest: widest Comb 3¼ Inches.
28 June	Lin'd the cockscomb-bed, which began to grow cool, with 9 barrows of very hot dung . . .
2 July	The best Comb five inches & half wide . . . Cockscomb-bed very hot with the new lining.
23 July	Took the Cockscombs out of their frame: the best comb full seven Inches wide; the leaves very large, & green; & the largest Stems two inches & a quarter round: the Combs well indented: That Amaranth that was suffer'd to run to many heads, looks very fine, & makes a pleasing variety. The wind is very apt to snap-off the leaves when the plants are first set-out, before the air has hardened them: heavy rains do the same. The tallest plant about three feet four inches. [To this is added, in the hand of Thomas White: 'Mem. the constant wet Weather rotted several of the Heads of those that stood abroad'.]
7 August	The best Combs grow mouldy.

The relative success of this culture (although Miller's *Dictionary*, 1768, claims 'plants five or six feet high, with crests near a foot in breadth')[16] encouraged experiment again in 1755, and on a hotbed 'two pots of Cockscombs' together with 'one pot of pendulous Amaranths' were sown on 15 March; and on 29 April there is the entry: 'Transplanted some Cockscomb-plants, not very forward, into one of the two-light melon-frames'. There, at least for this year, data about amaranths and cockscombs are concluded. Within a day or two White was to leave for Bristol (and a curacy in Wiltshire), and was not to make a further entry in the *Kalendar* until 17 July; in fact, no mention of his annuals occurs until 19 September, when there is the note that the 'double China-asters [*Callistephus* sp.] make a fine show'.[17] And cockscombs are not referred to again in the whole of the *Kalendar*.

What seems to have happened is this. Amaranths, in the form of love-lies-bleeding, were introduced to Wakes in 1752 by White's brother. White himself saw the possibility of the species and, at the first opportunity of a full season (1754), tried for success with the much more difficult, tender species, the cockscomb (cited by Abercrombie as a 'curious' plant in his 'First Class' annuals). As the entries show, White's enthusiasm was uninhibited and his culture meticulous, but some caution is hinted at: the season was not especially wet, yet the 7 August entry spells disappointment. In the following year a further opportunity was interrupted by clerical duty, and by 1756 there were other annuals, most notably the China asters, that were less demanding and perhaps, as the 19 September entry tells, equally showy. Interest in amaranths was not, however, exhausted, and the inclusion of *A. caudata* (White's pendulous amaranths) in the listing given above for 1761 annuals may be taken as representative for subsequent years also.

In contrast to this account of White's sudden and deep enthusiasm for growing cockscombs followed by an equally rapid loss of interest, a second example of the narrative rewards of the *Kalendar* is a story of slow growth, modestly expressed: indeed, unless we are alert as readers it is easy to think small potatoes of it. The first reference to *Solanum tuberosum* occurs in 1756, on 17 March: 'Planted some very large potatoes . . . The Ground was double-trench'd in the winter; & some rotten dung, & old thatch were dug-in at planting.' The outcome of this preparation is not recorded, and there is no mention of potatoes in the following year (1757), but on 28 March 1758 is the note: 'Planted 59 potatoes . . . not very large roots', and for the remaining years of the *Kalendar* there is a fairly regular record of planting and harvesting, with progressive attention to culture. For example, on 14 April 1759, when the potatoes were cut into pieces, we are given setting distances, and in the same year (November) detail of the yield:

> The potatoes, raised from about 14 large ones cut in pieces, turned-out a fine Crop of about 3 Bushels: several single ones weigh'd about a pound. Put-by about 30 of the finest as a supply for a crop next year.

The practice of saving a supply for seed-potatoes seems to have originated in this year, but is then carried forward; for example, a note on 26 March 1763 reads:

> Planted five rows of Potatoes quite across one of the middle quarters of the new-garden in well-dunged deep mould. The pieces were cut from large firm roots that had been well-preserved from ye frost. If the pieces had not been planted 15 inc: apart, they would not have held-out.

At about this time White became interested in the most effective way to manure the crop. As noted earlier (1756), he first used 'rotten dung, & old thatch'; by 1764 (22 March) he recorded that the 'ground had been well dunged, but no thatch was used', and in the following year (12 April) he initiated an experiment—with a control:

> Planted five rows of potatoes . . . & put old-thatch in four of the
> trenches, & peat-dust in one for experiment sake.

Unfortunately, the experiment was invalidated by modern standards
since the same entry continues: 'Exchang'd roots with Mr: Etty [at
Selborne vicarage], as his ground is so different: his sort came
originally from me',[18] but on 22 November White drew the conclusion:
'The potatoes turned-out well beyond expectation after such a burning summer: those planted on peat-dust were superior to those on old
thatch.' How consistently this culture was followed later cannot be
ascertained, but at least we know that 'peat-dust' was favoured 'in
every trench' the next year with the result that the crop was 'large, &
good' (9 April, 13 September 1766). What, of course, is interesting
here is both the experiment with growing conditions and the exchange
of seed: we know today that 'if potatoes are grown in England year
after year from seed saved from the crop grown on the same land in
the previous year, the stock degenerates and becomes extremely liable
to contract disease',[19] and it is heartening to find White's common
sense guiding him to good practice.

Within the context of the *Kalendar* little else can be said about
White's growing of potatoes, but if we move to something he wrote
much later the interest of the record is considerably heightened.
Writing under a date in 1778 (in a letter to Daines Barrington), White
remarks:

> As to the produce of a garden, every middle-aged person of observation
> may perceive, within his own memory, both in town and country, how
> vastly the consumption of vegetables is increased . . . Potatoes have
> prevailed in this little district [Selborne] . . . within these twenty years
> only; and are much esteemed here now by the poor, who would scarce
> have ventured to taste them in the last reign [that of George II
> 1727–60].[20]

In pronouncing thus, White confirms (although for a later date)
Miller's similar evidence from his *Dictionary*, where in 1768 he wrote:

> This plant [the potato] has been much propagated in England within
> thirty or forty years past . . . yet it was but little cultivated in England
> till of late; these roots being despised by the rich, and deemed only
> proper food for the meaner sort of persons; however, they are now
> generally esteemed by most people, and the quantity of them which are
> cultivated near London, I believe, exceeds that of any other part of
> Europe.

This claim, that the potato had been cultivated throughout the reign
of George II, offers a keen illustration of the relationship of a small

village in Hampshire to sound economy pursued in the metropolitan centre. It also demonstrates the role a village clergyman could play in promoting the welfare of ordinary folk: for, if it was White who introduced the potato to Selborne (and the record suggests very much that it was), not only did he continue to find the crop satisfying himself, even 'housing' it for winter use, but the villagers also learned to profit, White noting with satisfaction in autumn 1787:

> The quantity of potatoes planted in this parish was very great, & the produce, on ground unused to that root, prodigious. David Long had two hundred bushels on half an acre. Red or hog-potatoes are sold for six pence pr bushel.[21]

Moreover, the potato record confirms the interest that can be found in considering White's use of sites in the garden.[22] Although the ground used for potatoes is not mentioned every year, and although in some years the planting is made in the garden of a neighbour (Timothy Turner),[23] the record shows that White used a different station year by year: in one season potatoes are planted in a middle quarter of the new garden, in another at Turner's; 'a mellow rich part of the garden' is used, and later 'near the fruit-wall'. That detail of this kind is so often provided is not in itself persuasive, but taken in conjunction with other facets of White's writing it becomes forcibly clear that his attention to detail, accuracy of report, and ability to sustain a record over time were all extended throughout the period of the *Kalendar*; and enforce the judgement made earlier that the document can be read as a record of narrative particularity.

The rewards derived from analysis extend even further. That it is possible to ascertain what happens to this plant or that, at one or another site, hints in turn at other kinds of particularity. From where did White obtain his seeds and plants? What assistance did he have in the garden? And (and this is of crucial importance if the case is to be made that the *Kalendar* demonstrates a progressive and developmental acquisition of skill) is there anything in the *Kalendar* to help in understanding how White the country parson became correspondent with two metropolitan figures (Thomas Pennant and Daines Barrington, both of whom were Fellows of the Royal Society) about migration, bird song, instinct, and all the other various matters accomplished in *Selborne*?

To raise issues such as these moves discussion into the collaborative nature of White's work. In a very obvious way a glance at the early pages of the *Kalendar* shows that some of the entries (besides those mentioned above in connection with cockscombs) were made by hands other than White's. The extent of these hands is not great: they are confined to four years in the period 1751–55 (1753 is wholly White), and comprise in length no more than about 1 per cent of the whole *Kalendar*, that is, about four pages in all out of a document of over 400 pages. The entries made by these hands, belonging to four (possibly five) different people, all of whom were members of the family,[24] fall into two groups. In the one group are remarks that report routine activity in the garden (for example, the entries for seven days in October–November 1754 that mainly concern autumn sowing and planting of bulbs and perennial plants); in the other groups go single entries, reporting acts that are critical. These latter, eight in all, are of sufficient importance to be cited in full:

```
1754
15 June          – Cut first Cucumber . . .
 7 August        – Cut first Melon . . .
17 December      – Put the Spawn into the Mushroom Bed

1755
27 March         – Sowed more Melons in the Pots that fail'd.
31 March         – Sowed one pot of Mr Garniers Cantalupe 1753 . . .
23 June          – Cut the first Cucumber.
 1 August        – Cut the first Melon . . .
15 September     – Planted the mushroom spawn . . .
```

From the absence of entries made by White himself within a space of a few days before or after each of these items (and in some cases much longer, e.g. in 1755 the entry immediately prior to that of 23 June is dated 29 April, and the one after is 17 July, both in White's hand), it is clear that White himself was away from Selborne. The natural processes in a garden, however, do not wait upon the presence of a gardener: plants will flourish (and on occasion die) whether we will or not; and it seems clear that in the instances given White left instructions that certain occurrences were not to pass unrecorded.[25] It is evident that it mattered to him that the growing season for the first cucumber or melon could be ascertained: this is indicated not only by the exclusivity of what is noted, but also by the fact that White himself recorded similar details in other years. In addition, we know that the facts cited were used by him in later entries: for example, on 26 August 1755, 'Gather'd the first Mushrooms from spawn put into a bed last Decemr: ye 17th.', and in the following year (1756) there is the note on 2 August, 'Cut first Melon . . . very early, considering the first bed was destroy'd'. Taken on their own, such data for a past age appear of little consequence: only in recent years has it become possible to control growing conditions to ensure flowering and fruiting of a range of plants at a desired date,[26] and any records of the kind people such as White kept must seem of human and antiquarian interest only. In part, this is so; but the collection of scientific data about optimum sowing times for a target harvest had to begin somewhere, and in White's *Kalendar* we can see interest in a search that is still being extended in our own century. How White himself became actively participant in the collection of similar and more detailed records will be part of a later chapter: at this stage we can merely note, as with so many aspects of his published work, that the first moves towards data for which he was to become so well known (the natural history of his parish) are embedded in the records derived from work in his garden.

This evidence of records made by other hands is sufficient on its own to indicate that work in the garden was a collaborative undertaking involving many members of the family; but we know also that additional (paid) help was readily available, as well as regular assistance from White's gardeners, John Beckhurst and Thomas Hoar.[27] The various enterprises were dependent upon plants and seeds that came, for the most part, from outside the village. Some seed was of his own saving,[28] occasionally plants came from neighbours, and good use was made of what grew in the ditches, hedges, and other habitats in the parish.[29] The dominant impression from reading the *Kalendar* (confirmed by numerical counts) is, however, that the vast majority of White's sources were further afield, and, of course, this must be so: the community of keen gardeners with whom an enthusiast of White's skill and perseverance would be likely to exchange seeds and plants just did not exist in his immediate locality. This would seem to limit the opportunity for acquiring new material, but White made regular journeys to Oxford, London and other centres where relatives and friends lived, and all of them at one time or another furnished stock that was later to grace the borders (even sometimes the table) at Wakes. For instance, within a day's ride or less of Selborne was his uncle at Bradley, from whom he obtained gillyflowers and lilac suckers; at Bishop's Waltham was James Gibson, the future naval chaplain, who provided seed for peppers, white cucumber plants, pendulous amaranths and sweet williams for the borders, and who (after Quebec) gave White celeriac and 'Polyanth-seeds . . . said to be good'; from Mr Hunter at Waverley (Farnham) came laurels, seeds of a 'red-seeded Cantaleupe', and pine-strawberry plants; and from Sir Simeon Stuart he received the curiosity of 'a few Indian-turnep-seeds'.[30] Other sources for plants were his aunt at Ringmer, near Lewes; friends at Oxford (most influentially perhaps John Bosworth, a Fellow of Oriel); and his brother Thomas in London.[31] Indeed, Thomas is probably the most regular source of supply for the whole of White's gardening career, his gifts being assiduously noted throughout the *Kalendar*.

White also bought plants: at the time of stocking the fruit-wall, many of the trees came from Murdoch Middleton (although they were delivered very late in the season, on Boxing Day 1761); earlier (1756) he bought plants from Williamson's Nursery at Kensington, and on several occasions made use of a local nursery at North Warnborough, near Odiham, Basingstoke.[32] Indeed, the very first reference in the *Kalendar* to a source (24 January 1751) shows that White was prepared to spend money on the garden. After noting the sowing of radish, lettuce and onions, he continues: 'Planted-out five bulbs of the Crown Imperial (which I had from a Seedsman in London) . . .'[33] Yet it would be wrong to think that plants came only into Wakes: White himself also gave them away — and not just plants. On 13 February 1765 is the note: 'Sent-down a large portmanteau full of all sorts of perennials to my Brother Harry at Fifield [near Andover]'; on 3 April 1761 is the entry: 'Sowed a great Quantity of Cucumber-seeds for the neighbours'; and ten years earlier he had recorded on 7 March: 'Planted five young passion-flower plants, which I had from Oxon. Gave my U: White four'.

The range and extent of all these varied sources should not be regarded as unimportant. At a time when the dignitaries of Britain were competing for the services of Capability Brown to lay out their estates in the latest fashion, White and his peers were quietly gathering knowledge through the collaborative endeavours of searching out this plant and that, exchanging items and ideas, and generally sustaining the developing and rich history that is the English country garden. More specifically, during his experiments season after season, White came to recognise that for successful cultivation many plants require special conditions and attention,[34] and (in some cases) special equipment.

As indicated already, one of the main difficulties at Wakes was the clay soil. To counteract this, White made use of numerous materials, in addition to the regular purchase of dung to supplement that from his own horse.[35] The special equipment required is mentioned most frequently in connection with his melons and his cucumbers. The year 1756 brought a very wet spring and White records: 'The 20th [April] was a vast rain: but on the 26th: it rain'd for 22 Hours without ceasing, & brought on such a vast flood as has seldom been seen . . . My ten-light Cantaleupe-bed so flooded by those vast rains that all the plants are dead.' As a consequence some action was necessary, and at 3 January 1757 we read: 'Levelled, & widen'd the Area of ye

Melon-Ground; having made an underground Drain to prevent it's being flooded any more'; and the following year, on 17 January 1758, he took precaution over the condition of his compost: 'Finished an earth-house in the melon-ground. It is worked in a circular shape with rods & coped over with the same, & then well thatched: is nine feet over & eight feet high; & has room to hold a good Quantity of mould, & a man at work without any inconvenience.' As for attention to cucumbers, focus is on ventilation problems, and in February 1758 he attempted a remedy: 'Made a cucumber-bed full fourteen feet long, & almost four feet deep at the back for my two two-light frames . . . Laid a leaden-pipe into the frame that has got the tin-chimney, (according to Dr: Hales's proposal)[36] up thro' the back of the bed, in order to convey-in a succession of fresh air a nights.' This arrangement was tested the following month (25 March):

> Tryed an experiment late in ye evening with a Candle on the two Cucumber-frames after they had been close covered-up some Hours. On putting the Candle down a few Inches into that frame that has leaded lights & no Chimney, the flame was extinguished at once three several times by the foul vapour: while the frame with the tiled lights, & Chimney was so free from vapour that it had no sensible effect on the flame. I then applied the candle to the top of the Chimney, from whence issued so much steam as to affect the flame, tho' not put it out. Hence it is apparent that this Invention must be a benefit to plants in Hot-beds by preventing them from being stewed in the night time in the exhalations that arise from the dung, & yir own leaves. The melons confirm the matter, being unusually green & vigorous for their age. I applyed the Candle to the nose of the leaden pipe; but it had no effect on it: so that what air comes-up thro' it must be wholesome, & free from vapour.

Ventilation was not the only problem: protection (against frost and sun) was also necessary. So, one year (1751) he made 'a cover of oiled paper'; in 1757 he constructed 'a melon paper-House 8 feet long, & 5 feet wide: to be covered with the best writing paper'. Protection was afforded in another year to 'white Broccoli-seeds [which were] shaded . . . well with boughs'; and when the fruit trees on the wall began to blossom—in a spring (1766 for example) of cold nights but sunny days—we can read (20 March): 'Sheltered the wall-trees (which are too much blown) with boards, & doors'.[37]

Moreover, as he experimented, White came to appreciate the rich variety of the plant kingdom, grappled with some of its idiosyncrasies, and learned many of the skills that were to be of value later when to the plants of his garden were to be added the much more extensive flora and fauna of the parish. The rigour, and the curiosity, the patience and the wonder, all of which were to characterise his work as a naturalist, were born in the garden. As Senior Fellow of his college White would merit a footnote in the annals of Oriel, but it is primarily because of the opportunities that beckoned and were seized amidst the borders and frames, basons and quarters, of his garden that the world was later to greet as a gentleman and scientist someone imbued with the pastoral civilities and gracious benignities of an eighteenth-century Eden.

Before we move outside the garden, into the fields and ditches, the lanes and woods, there remain several further aspects of the *Kalendar* to note: they concern meteorology, experimentation, integrity, and observation. White's grappling with data about the weather figures most prominently in the future *Journal*, but if the opinion offered earlier—that it was White's brother-in-law, Thomas Barker, who prompted the keeping of the garden record—is to receive full acknowledgement it must also be known that Barker's interest in meteorology was something White espoused from 1751 on. At first, the notes in the *Kalendar* about the weather are seen from the vantage point of the garden alone. For instance, as early as March and April in 1751 White is recording conditions in this fashion: 'This bed [the hot-bed, made 21 March] by means of the great rains lost it's Heat; so that the Cucumbers, Melons, & Squashes never came-up'; and on 2 and 3 April: 'Planted four Asparagus-beds with plants of my own raising . . . sowed a thin Crop of Onions upon them. The Ground was well sanded, & trenched deep with good rotten Dung, but wet when planted.' The annual summary also is written from the perspective of the gardener:

> The Year 1751. was one of the wettest Years in the memory of Man. There were constant Storms, & Gluts of rain from the 20th: of Feb: to the 20th: of May. Part of May, & all June were very dry, & burning. But all July, & great part of August were as wet as ever: so that nothing in Gardens in a clayey soil grew to any size: & nothing came to bear 'till five or six weeks later than usual.

This focus on the relation between the weather and the garden is sustained throughout the *Kalendar*, and many of the entries show an important characteristic of early natural history observation: that is, of recording not the obvious but the unusual. A typical example from the middle years of the *Kalendar* is this, for 12 February 1759:

> Perfect summer: the air full of Gnats: & the surface of the Ground full of spiders webs, as in a fine day in August. The sun lay so hot on the frame that the Cucumber-plants wanted to be shaded.

Certainly from 1764 onwards, if not earlier, many of the entries begin to show an interest in the weather not just for its effects on the success or otherwise of the garden, but for itself alone as a phenomenon meriting the closest scrutiny. Consider the entry for 28 February 1765:

> A great snow with a fierce driving wind from the West, which forced it into every cranny & opening: so that the peat & mould in the houses were covered. It lies in very unequal depths on the Ground, being drifted by the strong wind: but would have been about ten inches in general had the air been still. The ever-greens were so loaded that they were weigh'd-down to the Ground. The wind was so strong, & the snow so searching, that the Hotbeds were not uncovered above two Hours all day. The sun broke-out in the evening: but ye Horizon looked very threatning, being of a very livid Colour, & promising more fall. The Mercury fell very low indeed in the night; & was quite concave at the top when I went to bed.

Admittedly, gardening concerns still figure here but, alongside, there is a new dynamic to the writing: a concern to find a vocabulary of descriptive, objective realities that focuses attention more on the meteorological matters than on the horticultural. It is a concern that will become extended and refined later. One further characteristic of this new writing should also be recognised: an interest in measurement. For us today, in an age of generally accepted standard units, it is difficult to grasp the slow steps by which people as educated as White came to contribute to the need for standardisation. True, in the early years of the *Kalendar* linear measure is readily available to ascertain the size of the cockscombs, but circumstances often preclude the ready availability of an appropriate instrument; and other quantities need measuring as well as what will fit into feet and inches. In such circumstances White instinctively sought the natural measures of his daily round: for instance, succades are referred to as the size of pigeon eggs, and when many ripen rapidly they 'come by Heaps' (12 June, 25 July 1765); the year before, particular specimens of the same fruit are named 'as large as Goose-eggs' (2 June); and in 1760 (7 July), of the cantaloups that were set 'the biggest fruit [was] about the size of a hen's egg'. Equally fresh and immediate in conveying a readily graspable measure are some of the references for length and depth: in early April a cucumber is 'about as big as the top of one's finger'; in a dry season, 'cracks in the ground [are] deeper than ye length of a walking-stick'; and after a snowstorm, and even a partial thaw, the snow proves 'deeper than an Horse's belly'. At times, such a mode of reference becomes almost poetic: a field has been ploughed several times, and although the 'weeds are all kill'd, & the soil is baked as hard as a stone; & is as rough as the sea in an hard Gale: the Clods stand [on] end as high as one's knees' (13 July 1765). It would be wrong, though, to think such measures atypical of the work of a practical scientist in the field; after all, in our own time craftsmen and others continue to deploy the deceptively simple to accomplish tasks that would often seem to require precise instrumental guidance,[38] just as in White's historic past men as competent as Galileo measured time by the beats of a human heart or by the dripping of water through a hole in a bucket.[39]

Equally homely are some of the experiments White engaged in. In a passage given earlier, mention is made of successfully removing 'a mouldiness' from some melon plants by 'the experiment of clipping-off the infected part with a pair of scissors'; and note has been made of experimenting with peat dust instead of old thatch in the bottom of the potato trenches. These are not all. On 24 January 1759 is the entry: 'Sowed a pot with Cucumber-seeds, & set it by the parlour-fire for experiment-sake', with a later note attached: 'These seeds came-up, but would not advance beyond the two first leaves' — (the same trial is not recorded again); the complex and detailed experiment carried out in connection with ventilating a hotbed has already been described; some melon plants are stopped 'very short towards the bottom of the runners, for experiment sake, to see what the small wood about the stems will do'; and the spirit of enquiry extends even to brewing, for in one year, when preparing 'half Hogsh: of strong-beer', White chose to use 'only rain-water [instead of water from the well] to try the difference'.[40] The results of this experiment were initially inconclusive; the following year water from the well was again eschewed, but it was used in a later year (1771) and on several occasions subsequently.[41] That White's procedures in some of these fields are variable should not disturb. Our focus must lie more in the recognition that he embraced experiment and was not tied to methods laid down by family (or other) tradition. The necessity for sustained consistency was not yet upon him; and when it arrived in 1768 with the format of the *Naturalist's Journal*, his apparently random approaches would neither inhibit nor challenge the amassing of data that is remarkable for its thoroughness.

To raise the notion of how thorough White's records are must also prompt enquiry about their integrity: how accurate are the records White made? A full answer to this must come later, during analysis of the *Journal*, but it is worth at this stage noting related questions in two independent areas. One is concerned with actual observation; the other with the keeping of records. When enquiring about the validity of what White writes, we need to establish a view of how trustworthy his identifications are (this is especially important in connection with observations in the field). We need also to seek an answer to a number of questions about the written records themselves. That a man makes accurate garden and field observations is no guarantee that the subsequent writing up of these data is also accurate. Further, inspection of the variety of records that will be made in the *Journal* must prompt caution about reliability across such varied areas: if accuracy is established in one area, does that necessarily mean a like accuracy can be relied upon in every other area? It may seem surprising to raise issues of this kind so early in a study of White, but what we need to grasp is that many of the characteristics of White's procedures and writing in *Selborne* were already well established by the time he came to compose the work that made him famous. In fact, they began to become an integral part of his behaviour when he first studied natural history seriously, and that time was when he was writing the *Kalendar*. The important question of White's veracity can therefore be partially answered on the basis of what we have found already in presenting a view of the garden records.

Glancing at any representative page of the *Kalendar*, one is struck by the confidence of White's hand, and by the assurance of the writing: for page after page, year after year, the security of certainty seems to stride the paper (see Plate 3). At first, such impeccable pages almost prompt disbelief: the document was a working document, not a fair-copy made up from notes, and a working document written with quill pens for family use and record rather than for public display. Gradually, however, any sense of disbelief induced by the visual appeal of the manuscript gives way to recognition of a compelling integrity. Just now and then one's eye is caught by a careful correction. In some cases these are for grammar or style and need not detain us, but occasionally they are for accuracy. For instance, after writing one spring a 'vast white hoar-frost on the Grass', his pen returns and deletes 'white'; at another time, in fact on the very first page of the document, there is a correction as to which row of celery has been earthed up, 'the last row' being changed to 'a row'; and in reporting the measured girth of a tree a correction is made from seven feet five and one-half inches to just seven feet five inches. Elsewhere, 'earth' is changed to 'dung', and 'wounded' to 'severed'. In the sum these changes are few; but they are significant pointers to a strict care for accuracy, and we can read White's records with greater security for having noted them.[42]

Finally, what does the *Kalendar* say, not about the garden specifically, but about natural history? Contrary perhaps to what one might expect from the full title, the *Garden-Kalendar* includes numerous entries relating to observations of birds, insects, plants and other phenomena, but consideration of this early evidence of White's growing interest in these areas will be delayed until his life has been brought forward to the time of his serious study of such things in the mid 1760s.

CHAPTER SIX

'Lights & Shades of ye Prospect':
1754–1765

Fêted, we might say, by the accomplishments and endeavours of the *Kalendar,* it would be easy to conclude that White's garden commanded his full attention and left little opportunity of either time or imagination for other pursuits. On the evidence of the *Kalendar* alone such a conclusion would be inevitable, especially since the document is confined almost exclusively to gardening concerns.[1] Yet, to have included in a dated record of the garden details of a personal nature would have been intrusive and a betrayal of an eighteenth-century understanding of literary appropriateness. Let us, then, acknowledge a similar decorum by leaving the garden in the rich splendours of the improvements achieved by the early 1760s in order to bring White himself, from where we last saw him professionally (as curate at Durley in autumn 1753), to the time (1763 and beyond) when, as Mulso put it, he had 'arrived at that happiest of human States, Independence', having 'all in [his] own Power, without a Necessity of attending & solliciting any Body'.[2]

Life at Durley was essentially rural and unrestricting; and White embraced it with alacrity. The duties were light: service at church once on Sunday with a sermon, once without, and the occasional sacrament, were all that was required.[3] Naturally, there were the gentry to entertain (twice yearly was what White paid for),[4] and local people to talk with about country matters; but otherwise the time was his own. He maintained an interest in the classics (Mulso says he translated from Horace), and thought about his sermons (his accounts show that he bought Joseph Butler's *Sermons* (1726) and *Analogy of Religion* (1736)); he attended to his garden at Wakes, received regular news from Mulso about his friend's activities and anxieties (especially the threatened war with France), and hunted (particularly partridge). He also travelled, and made visits. In the latter months of 1753, his accounts show that he was at Southampton and Oxford, as well as at several of the villages near to Durley; and in the next year, in addition to occasions at Alresford, Alton, Botley, Bradley (which he visited regularly since his uncle was vicar there), Soberton, Swarraton, Farnham (to buy trees for Wakes), and

Basingstoke (to the races), he was again at Southampton, spent time in Chichester (where he watched needles being made), visited Mulso at Sunbury and his brothers in London (where he bought a mahogany table for his uncle), and in the autumn spent a month at Oriel.

As a Fellow of his college, White could be called to Oxford at any time, although customarily college affairs were conducted at Eastertide. In this instance (1754) two matters required his attendance: the appointment of a vicar at St Mary's (Oxford) consequent upon the death of a Mr Whiting, and the academic prospects of his youngest brother, Henry, who was a candidate for a post-graduate exhibition at Oriel, both the new vicar and the new exhibitioner being determined by the college.[5]

That elections of this kind were fraught with politics (potential and real) within and without the college is shown by contemporary comment. In the case of the Bishop Robinson Exhibitioner, White so successfully canvassed support for his brother, who was elected early in November, that Mulso remarked (playfully, surely?) that he had thereby 'establish'd [himself as something] . . . of a Plotter'.[6] The affair of St Mary's, however, was altogether more important, since it affords a glimpse of matters central to White's future: politics.

The political scene affecting college elections and patronage more generally at this period was far more complex than it might seem; and to steer a clear course through the machinations, bribery and pressures of the times required more from White than he was able to give. In outline the various sympathies can be grasped easily, but the detail is often forbidding. The background to White's position was this: at the time of the Hanoverian succession in 1714, Oxford—where support in the previous century for the Stuarts had been very strong—greeted the new reign with considerable misgiving. In essence, both the university and the Church of England (which looked to Oxford to provide the intellectual arguments for its place in government) believed their authority in the affairs of state derived from God himself, and hence saw the pragmatic and 'mechanistic doctrines [of the new age as] a betrayal of all things sacred'.[7] The voice of the university was not, however, unanimous (there were Whigs as well as Tories, court politicians as well as church, Hanoverians as well as supporters of the Stuart cause, high-church Whigs as well as country politicians . . .), and White's own college, Oriel, was no exception.

Early in the century, Dr George Carter (Provost at Oriel) had first fawned on the Tory administration of Queen Anne, and then on the government of George I (profiting from both). As a head of college his powers in the election of fellows had been substantial, and by mid-century, although the traditional country (clerical and Tory) strengths of the college had been re-asserted, there still remained the sour taste and bitter residue of Carter's cringing self-interest. At the time White travelled to Oriel consequent upon Whiting's death, Mulso heartily hoped

> that All Party Rage had died with [Whiting]; & that not so much because we have increasing Obligations to the present Family, but for a real Regard for the University, which is in a very low Consideration for the Sake of a Parcell of Fools who are a Disgrace to it in every View, & are of the most contracted Hearts of any Set of Men that I know.[8]

Whether White reciprocated these views is not recorded, but the political dissension alluded to must have distressed a man of his temper.[9] Such dissension as still existed neither diminished his loyalty to the college nor extinguished finally his earlier hopes of academic preferment—although, as we shall see, his own inclinations were to obstruct his election not only to a rich college living (and the prospect of complete independence, and perhaps marriage), but also to clerical security within the neighbourhood of his own birth and upbringing. But this is to anticipate.

After joining in the necessary discussions at Oxford in autumn 1754, White returned to Hampshire and duties at Durley. Except for the death of his grandmother (in her early nineties), the winter passed with its customary visits and activities, but in the following spring (1755) a new opportunity arose: to be curate of a parish in Wiltshire, West Dene. The vicar, the Reverend Edmund Yalden, held office also at Newton Valence, a parish adjacent to Selborne, and the family had already established several long-standing relationships with the Whites.[10] It must have seemed a sensible arrangement. White visited West Dene with Mr Yalden 5–8 March and agreed to accept the position.[11] In the summer he spent a second season at the Hot Wells in Bristol; and with the approach of autumn discharged duty at West Dene—for example, he conducted a wedding there in August, and later read banns on four consecutive Sundays.[12] But there was ample time for other pursuits: he learned something about the finding of truffles (the discovery of which in the Selborne area he was later to record assiduously in the *Journal*),[13] and received visitors. Thomas Barker, his brother-in-law from Rutland, was at Selborne during August and September and evidently accompanied White to Stonehenge, to Salisbury and to Wilton House, where the horn-room proved a lasting image since White was to refer to it in *Selborne* thirty years hence.[14]

However, appropriate as this kind of life must have seemed to a man with White's interests (for he was able to continue to hunt, partridge and hare especially, and Selborne was near enough for him to maintain an active interest in the garden), he missed the company and reassurance of his home locality, and Mulso wrote in September 1755:

> I hope [this letter] will find You well, & reconciled to your Situation; which, tho' you have as much true Philosophy as any Man I know, yet is not to your Taste, if it is really solitary. You have I suppose by this Time made a small Acquaintance about You, & perhaps have a Friend or two who will help to scrape your Blade Bone, after clearing the way to it at Dinner.[15]

Neither the anticipated reconciliation nor the friends, however, were forthcoming, and White soon accepted duties at Newton Valence (Mr Yalden's premier parish), which he was able to serve from Selborne. As it happened, this move marked the final stage in White's clerical peregrinations. Although in future years he was to hold other curacies, most notably one in Northamptonshire (where he was able to be non-resident), one at Faringdon (a parish on the northwest boundary of Selborne), and one at Selborne itself, it was this last, his native village, that was to be his emotional and physical home for the remainder of his life.

The desire to live permanently at Selborne provided another satisfaction: the opportunity to care more attentively for his father. John White had been in declining health for some months, and must have welcomed the more regular presence of his eldest child. From later evidence we know that White took his responsibilities as head of the family seriously, and his recording in the *Kalendar* (see Chapter 2) plantings designed particularly for his father's pleasure is a moving indication of his filial regard and gentleness.

Settled as he now seemed, it comes as something of a surprise to find that in the winter of the following year (1756/7) White contemplated permanent office away from Selborne—at Oriel. The successor to Dr Carter as Provost, Walter Hodges, died in January 1757, and White decided to offer himself as a candidate for election. From this distance of time the decision, late in the making, and prompted perhaps as much by his collegiate seniority and the hope of a secure financial future as by the wish to serve the college, seems unwise and distinctly lacking any prospect of success. True, the college had accepted his precedence in 1752 for the proctorship, but on such a flimsy basis to decide, even with the support of friends, to put himself forward for provost suggests a limited appreciation of what such a post involved. The constituency was always the college Fellows, and their duty was plain: to further the dignity and purposes of the college. In many instances this meant that the election was made of heads who had the resources to contribute to the material worth of the college either in their lifetime or by testamentary provision, or who were expected to hold professorial status or obtain preferment to high office in the church. White's assets, in any of the necessary categories, were insubstantial: his family held no great estate; his influence, whether considered from the perspective of the court, of the government, or of the church, was slight;[16] and, although he continued to read the classics and to write some poetry, his intellectual powers were unexceptional. What he did have, of course, was a secure understanding of life in a country parish, an established preference for residence out of college, an interest in field diversions, and an enthusiasm for the natural scene. Even with a genial manner, however, such characteristics are scarcely those one would seek in a provost, and not unexpectedly he lost the election. His successful rival, Chardin Musgrave, proved magnanimous in victory and in autumn of the same year White was offered, and, after discussion of terms that would enable him to be non-resident, accepted, the living of Moreton Pinkney in Northamptonshire.[17] Although in later years he was offered several other livings, any of which would have brought a considerably higher income, Moreton Pinkney was the only preferment he accepted from Oriel, and he held it until his death.

That White would choose to decline further preferment with an enhanced income is initially puzzling: he was never sybaritic, nor is there any suggestion that he favoured an ascetic existence. The answer must lie in the sum of several factors, but of two in particular: first and most notably the question of his patrimony, and secondly the terms on which a fellowship was held—both of these going some way to explain his lifelong bachelorhood.

In the course of 1757 further duty was carried out at West Dene and Newton Valence[18] and, consequent upon the ill-health of the vicar, Dr Bristowe, at Selborne itself; this last duty White also resumed in 1758 when the vicar died. But the loss of a vicar was not the only death sustained that autumn at Selborne, for on 29 September White's father died—to be buried simply, without stone or other memorial, wishing as his will states to leave this world 'with as little show and expense as may be and . . . not desiring to have my name recorded save in the Book of Life'.[19]

News of this event was not long in reaching Oxford and it was generally expected that White would resign his fellowship within the year. Such an expectation may seem strange, but the circumstances in which fellowships were held in White's day were different from those of today. As Mulso was to remind his friend:

> Fellowships are a Sort of temporary Establishments for men of good
> Learning and small Fortunes, 'till their Merits or some fortunate Turn
> pushes them into ye World, and enables them to relinquish to Men
> under the same Predicament.[20]

The kinds of 'fortunate Turn' Mulso had in mind were marriage to a
rich heiress, preferment to what he termed 'Fat Goose' living (a parish
incumbency with valuable tithes and good glebe) or, as in White's
case, a family inheritance. But the facts of White's circumstances were
not what the world surmised: John White had little to leave his eldest
son, and certainly less than would force a man to resign a fellowship.[21]
White knew the justice of his case and determined, if necessary, to
appeal to the Visitor of the college to ensure that his position was fairly
dealt with. Fortunately, such a dire step was not needed. After some
correspondence and discussion with Musgrave, who pronounced that
'it was in [White's] own breast to keep or leave [his] Fellowship; for
Nobody meant to turn [him] out', White was reconciled to his
position and congratulated by Mulso for being no longer 'troubled
wth Party & Contention', an outcome he attributed to his friend's
poetic placidity of mind.[22]

Security now in the retention of his fellowship, White's prospects,
unworldly though they might seem, were assured, and he was able to
withdraw quietly into his home and parish. In late summer Mulso
and his wife came to stay at Selborne, and in November the same year
(1759) White set off with his sister for the longest absence from
Selborne he was ever to know. Calling first at Sunbury (to see the
Mulsos), he then stayed in London with brother Thomas, and early
in the spring of 1760 set off north to visit sister Anne and Thomas
Barker at Lyndon. He was not to return to Selborne until 17 May,[23]
and was to be away for almost precisely six months, after which he
found his garden

> in very good order, considering the long drouth this spring.
> The Cucumbers in full bearing . . . The Cantaleupe-melons in good
> Condition, & just shewing fruit; & the Succades very stocky plants . . .
> All the kitchen-crops . . . in good plight: & the Coss, & hardy lettuce
> that stood the winter, very fine.[24]

White's devotion to his garden has already been discussed, but before
turning to his more general interest in natural history we must attend
to three additional matters concerning his personal affairs: ownership
of Wakes, the prospect of succeeding his uncle Charles White as vicar

of Bradley (which could be served from Selborne), and the possibility of marriage.

This last has often been mentioned by White's commentators; and, whilst it is true that he was regularly chafed at one period by Mulso about the advantages of marriage (prompted in part by White's own speculation), there is much to suggest he was a confirmed bachelor.[25] Ostensibly, there is an argument that suggests that his resources both before and after his father died, even when supplemented by the income from his curacies, were insufficient to provide for a bride, at least in the fashion he would wish. The only way to extend his financial prosperity would have been to accept one of the Oriel livings that were offered him fairly frequently.[26] This in turn would, of course, have necessitated resigning his fellowship; and so, incidentally, would marriage. Circumstances, then, were against his taking a bride; and so perhaps were affairs at home. At the very moment he became, one might say, his own man at Wakes (upon the deaths of his matriarchal grandmother and his father), he thereby assumed increased responsibility as head of the family—but had no settled income. One possibility seemed to remain. Of all the livings that were open to him that could be served from Selborne (which, as the family home, with his brothers and sisters now nearly all flown, he had surely to maintain), the only one that might fall to him would be his uncle's at Bradley. This was in the gift of Robert Henley, the Lord Chancellor, who lived at the Grange, near Alresford, and to whom he occasionally sent the gift of a choice melon.[27]

Opportunity to test the Chancellor's beneficence arose in 1763. For the past year or so, White had accepted duty at Faringdon, as curate following the death of Stephen Hales; his brother Henry had accepted the living of Fyfield, near Andover (and there was the possibility of further patronage from the Lord Chancellor); and now (in March 1763) his uncle had died. One result of this was that White became the outright owner of Wakes and no longer paid rent for the patrimonial home; but, more importantly, Bradley was vacant. He applied to the Lord Chancellor, and received a rapid rebuff. The reason for this had been chiefly determined some years before when, at the election of a new Chancellor for Oxford, White had voted with some reluctance for the candidate supported by the government (his natural inclination being to vote for the Jacobite candidate).[28] Such disloyalty had been quietly noted, and the end of hopes for preferment to a living that could be served from Selborne was the outcome.

Although this decision must have reached White as a shock (in one blow it dashed his hopes of clerical security and marriage, both tied to his native territory), it may not have been totally unfortunate; and his natural delicacy of feeling and fastidiousness of manner may actually have been rapidly reconciled to these new circumstances. Selborne, rich in 'nature's rude magnificence', provided a 'rural, shelter'd, unobserv'd retreat' and a fitting consort;[29] any other might have proved a distraction greater than he could afford.

Having arrived at that independence, however modest, which Mulso termed the 'happiest of human States', one might surmise that White's cup was full. He still, however, thought about a college living and, if Mulso is to be believed, marriage.[30] In the short term there were plenty of things to occupy his time and mind; and visits to make. There were also guests to receive[31] and the garden to tend. The restlessness of his condition was evident, as was a certain melancholy; and yet, although much was done, there was no opportunity to do what his friend most wished—visit him in the Yorkshire parish to which he had moved in 1760. In fact, Mulso characterised White at this time (1765) with great percipience:

> Your Description of your sitting in your dining room reminds me of your old Situation . . . For Shame, Gil! this Vacuity ought to have been filled up . . . Five years gone! & a Batchelor, & find no *Time!*[32]

Indeed, there *was* no time to visit Yorkshire. The next stage in White's life had arrived; and the period of vacuity was to be replaced by a thorough-going absorption in a new field: work on the natural calendar.

CHAPTER SEVEN

The Natural Calendar and
Flora Selborniensis

It was remarked earlier that, although the *Kalendar* provided White with the first opportunity to record data about the natural world, his serious study of natural history did not begin until the mid-1760s. The main evidence for this is the document known as *Flora Selborniensis*, which comprises a detailed account of observations, principally at Selborne but elsewhere also, during the year 1766.[1] Typically for White, the document is not single-minded; it includes much more than a listing of the flowers of the parish, and this width of interest is announced by the sub-title: 'Flora Selborniensis: with some co-incidences of the coming, & departure of birds of passage, & insects; & the appearing of Reptiles'. Even this enlarged title, however, is not fully comprehensive, for the *Flora* also includes occasional entries about weather conditions. Two hundred years after its composition, it is easy to be misled by White's use of the term 'co-incidences'. For us, the word is normally used to indicate the chance occurrence of events happening at the same time; yet, if we are to understand the full purpose of both the *Flora* and the later *Journal*, we must grasp the quite different indications of usage current in the eighteenth century.

The key text is a volume published by Benjamin Stillingfleet termed *Miscellaneous Tracts relating to Natural History, Husbandry, and Physick*.[2] To the second edition of this work (1762) Stillingfleet added a calendar prepared by Alexander Berger at Uppsala in Sweden in 1755, and accompanied his translation of Berger's work by a calendar of his own, prepared as it happened the same year as Berger's, but at Stratton in Norfolk. Each calendar is a bare listing throughout the year of natural occurrences, with a few notes: the items recorded are mostly plants, but occasional mention is made of birds and other creatures and some observations are made on the weather, though these are infrequent. The styles of each document are broadly similar, and the scope is best seen in two successive entries from Stillingfleet's calendar: they are taken from the record for the month of June.

19. *Thermom. 44.25. Highest this month.*
21. Orache, *wild*, 154.1. Chenopodium *album*, F.
    Solstice. About this time ROOKS *come not to their nest trees at night.*

Wheat, 386.1. Triticum *hybernum*, F.
RYE, 388.1. Secale *hybernum*, F.
Self-heal, 238. Prunella *vulgaris*, f.
Parsley, *hedge*, 219.4. Tordylium *anthriscus*, f.
Grasses of many kinds, as festuca, aira, agrostis, phleum cynosurus, in ear.[3]

Bare as this listing is, and (as it must seem) empty of any purpose other than as a bald record of plants with 'flowers full blown' ('F') or 'flowers beginning to open' ('f'), the preparation of such a calendar was attended by serious scientific purpose: not only the surveying and classification of the whole of the natural world, but the improvement of the quality of daily life for all people.

To realise how this might be, we need momentarily to remind ourselves that at the time Linnaeus made these proposals (for, in the sense we will explore them, they derive from him) there was no food that was ever canned, frozen or made up in the laboratory: everything (fruit, vegetables and other staples—for example, corn for bread and malt for beer) had to be freshly grown, simply dried, or salted. Yields, and hence prices, varied enormously from one season to another and, with fixed wage structures, created periods of great distress and misery, especially among the expanding urban populations.[4] In these circumstances, anything that could be done to improve farming methods and increase the yield of crops would better man's lot. The kinds of improvement advanced in England are familiar: Jethro Tull's seed drill of 1731 and the impetus given in the last quarter of the century by Arthur Young and William Marshall to the need for detailed records of local practice are, along with the establishment of the Board of Agriculture in 1793, amongst the most influential.[5] The basis of these endeavours was, of course, a desire to place husbandry on a rational basis, such that there were available to every farmer in the land principles of practice that were certain, year after year, to ensure success.

The listings in the Stillingfleet calendar must seem a far cry from such a high purpose, but, because the scale of what was advanced was so immense, there were opportunities for very varied contributions to its attainment. Whereas some for example were to improve yields by experimenting with soil preparation, others concentrated on the management of livestock, or the discovery of new varieties of plant. The contribution projected by Stillingfleet was more ambitious than these: he hoped to discover a secret of nature herself—when should a crop be sown, and when should it be reaped? Was there a natural calendar that could guide man's rural economy?

To ascertain this calendar, guidance it was thought might properly be sought in the natural world itself, from the leafing, flowering and fruiting of flowers and plants, and from the behaviour of birds, fish and other creatures. More significantly, although the occurrence of natural phenomena as single events might be observed annually, it was envisaged that careful records would show two patterns of correlation: first, that there were 'co-incidences' of occurrence in the natural world; second, that, once certain coincidences were known, careful observation would show that they were prognostic and could be taken as a natural guide to husbandry. Stillingfleet, wryly observing that 'when artificial calendars came into vogue the natural calendar seems to have been totally neglected', put it like this:

> I think we may assert universally, that whenever two things, however disparate in their nature, constantly accompany one another, they are both actuated and influenced by the same cause. Now that cause may probably operate on other things that lye within the reach of our powers, and depend on our determination. Thus that constitution of the air, which causes the cuckow to appear about the time, when the fig-tree puts forth its fruit, may indicate the properest season to sow some of our most useful seeds, or do some other work which it imports us to do at a right time; and that time may not be according to certain calendar days, but according to a hitherto unobserved calendar, which varies several weeks in different years. . . .[6]

There, in each of its main components—meteorological, botanical, ornithological—is the framework for Stillingfleet's calendar, and for White's too. The enormous scale and sheer labour of the task envisaged were never fully realised. Stillingfleet knew that what he proposed was only a preliminary to a more selective outcome: 'If ever any use [is to] be made of Calendars of this kind, it must be by finding out, after a long series of observations, and publishing by itself a list of a few regularly prognostic plants.'[7] Further, and this must have been especially appealing to White, the benefits that were expected to accrue to mankind would not be confined to agricultural husbandry. Others would profit also:

> If the gentlemen of our own, or other countries, took delight in such observations, they might amuse themselves very agreeably . . . as it might furnish materials for directing private æconomy, and the more so as the times for sowing of seeds, for reaping, and mowing, and for gathering fruits of various kinds . . .
>
> Gardeners might thence learn at what time of the spring, they ought to lay the roots of plants bare, when to sow their seeds, when to expose to the open air, and when to put under shelter their tender plants, and how to furnish the garden with flowering plants; so that there might be a perpetual blow all possible months of the year; thus the *lilac* follows

the *cherry*, the *mock orange* follows the *lilac*, and the *late roses* follow the *mock orange*.[8]

And lastly (again surely to White's satisfaction), the programme envisaged was Linnaean in scale yet firmly practical. In contrast to so much earlier literature about nature, which had relied upon hearsay and superstition and had thereby merely passed on the errors of ignorance, this new method of proceeding was to be grounded in first-hand experience. It was a mode strikingly in harmony with White's own inclinations, and one that accorded well with his enthusiasm, not for travelling by coach, but for riding on horseback and observing the natural scene directly. As he was to remark in *Selborne*, many '*Faunists* . . . [were] too apt to acquiesce in bare descriptions, and a few synonyms . . . [which] may be done at home in a man's study', a procedure in sharp contrast with what he himself professed: to be 'an *out-door naturalist*, one that takes his observations from the subject itself, and not from the writings of others'.[9] It was a declaration wholly within the post-Baconian and inductive, British empirical tradition, and one that later was to contribute strongly to the appeal of his writing to Anglo-Saxon readers. Before we approach the *Flora* and inspect the outcomes of his observations, however, it is worth commenting on how White came to embrace Stillingfleet and thereby initiate his own contribution to the Linnaean programme.

The *Garden-Kalendar* is indubitably about the garden; yet, as we have seen, other material creeps in. If we look carefully at the *Kalendar* records, we see that the first observations unrelated (in a narrow sense) to horticultural and meteorological concerns appear in 1758. On 9 April that year is the note, 'Saw two Swallows: one was seen in ye village on the 3rd'; and on 2 November is the longer entry:

> Saw a very unusual sight; a large flock of House-Martens playing about between our fields, & the Hanger. I never saw any of the swallow-kind later than the old 10: of Octobr: The Hanger being quite naked of leaves made the sight the more extraordinary.
>
> Warm wet weather for many days, with blowing nights, & sunny mornings.
>
> The leaf fallen more than usual.

In subsequent years occasional notes about birds also occur (for example, 4 May 1759, 'first Redstart, & Cherrysucker'; and 8 June 1761, 'Rooks are perchers'), but the incidence of such notes is low; and, except for a long—and therefore atypical—account of May 1761 concerning investigation with his brother Thomas of a colony of field crickets, *Gryllus campestris*, there is no substantial sequence of natural history observations until summer 1765. Then something happens.

Natural history observations do not just creep in, they positively erupt:

12 July: The Swallows & martins are bringing-out their young. Young partridges that were flyers seen.

14 July: Saw Pheasants that were flyers.

16 July: Several fern-owls or Goat-suckers flying about in the evening at Black-down House.

21 July: The Glow-worms no longer shine on the Common: in June they were very frequent. I once saw them twinkle in the South hams of Devon as late as the middle of Septemr:
The Redbreast just essays to sing.
Dry dark weather with an high glass.

23 July: The field-crickets cry yet faintly. Hot dry weather still.

28 July: The Martins begin to assemble round the weather-cock; & the Swallows on the wallnut-trees.
Dry hot weather still, with a N: wind.
The Goldfinch, Yellow-hammer, & sky-lark are the only birds that continue to sing. The red-breast is just beginning. The field-crickets in the Lythe cry no longer.

29 July: The beetles begin to hum about at the Close of day.

1 Aug: No rain at all since the 16 of July.

3 Aug: A plentiful rain for five hours & an half with a great deal of thunder & lightening. It soaked things thoro'ly to the roots, & fill'd many ponds.

4 Aug: The swifts have disappeared for several days.

5 Aug: Did a great stroke of Gardening after the rain . . .

7 Aug: Dripping warm weather since the thunder-storm.

8 Aug: Mr: Yalden saw a single swift. Glow-worms appear'd again pretty frequent; but more in the Hedges, & bushes than in June, when they were out on ye turf.

9 Aug: Saw two swallows feeding five young ones that had just left their nest: they usually bring them out the beginning of July.

Selective as these entries are (I omit some of the observations made on the weather, and all those about the garden), the sustained introduction into the *Kalendar* of observations about natural history is strikingly new. And although the knowledge that enabled White to make the entries (particularly those about birds) had been developing over many years — his earliest records go back to 1736[10] — the urge to make regular written notes was clearly a fresh departure. In fact, it heralded two further developments: an excursus into the mastery of formal botanical nomenclature, and the writing of the *Flora* itself.

As regards the first of these developments, it is usual for writers to remark that White's serious study of botany began in 1765.[11] That is true — as far as it goes.[12] Of far greater relevance, especially for a man whose first and last love in the natural scene was birds, is to discover why he ever decided to take up botanical study, and to enquire what benefits he expected to obtain. To approach these last, we need to recall what stage White's life had reached. His country childhood had been followed by Oxford, and then by ordination; in the next ten or so years hopes of an academic career at the university had been dashed, as had the chance to inherit his uncle's living at Bradley. These (and other) disappointments must have led to considerable self-doubt, and in such circumstances it is natural that he should have retreated to the quiet felicities of Selborne. Once there, he began to extend his library. He bought both travel literature and some of the standard natural history texts of his day;[13] and some time in 1764–5 he studied Stillingfleet's *Tracts*. The exact date on which he obtained the *Tracts* appears to be unrecorded, but by the summer of 1765 he felt so excited with what was evidently a recent acquisition that he urged his enthusiasm in a letter to Mulso of 2 August that year.[14] It acted as a catalyst, offering him an opportunity that must have seemed almost providential: of uniting in a single activity his public role as clergyman with his private passion for the natural world. By means of this one book, he was shown how, in a practical fashion, he might contribute to the natural theology to which he had been introduced as a schoolboy and which held that a beneficent Providence was present in the intricacies of Nature, and that scientific enquiry would reveal, in ever-increasing detail, 'the all-wise disposition of the Creator'.[15]

Before he was in a position to engage in this specific task, White knew that considerable learning would have to be accomplished. As the calendars of Stillingfleet (at Stratton in Norfolk) and of Berger (at Uppsala in Sweden) had shown, the guiding signals on which a natural calendar was to be founded were the leafing and flowering of plants. If a calendar was to be made for Selborne, a sound knowledge of botanical names, and of locations in the parish, would have to be secured; otherwise those observations he was already competent to make (about birds) would have no accompanying botanic 'co-inci-dences',[16] and the value of any calendar that might ensue would be defective at birth.

To prepare, then, for the detailed knowledge that would have to be deployed the following year (1766) — in all reasonableness the calendar would have to run for a full calendar year and begin on 1 January — White launched himself into botanical study. The word 'launched' suggests perhaps something more immediately dramatic than what actually happened, for the beginnings of study were decidedly tentative. Although the *Kalendar* records at about 9 August 1765[17] the sharply technical distinctions by which *Scirpus* sp. may be distinguished from *Juncus* sp., any kind of follow-up to this initial deployment of technical terms is delayed for almost a fortnight, at which time (22 August) White noted:

> Wild-ragwort, scabiouss, hawkweed, knap-weed, burdock, yarrow, rest-harrow, &c: in flower.

The ordinariness of the list is immediately apparent, but a study has to begin somewhere and what might well have been found within the boundary of Wakes itself is as good a place as anywhere else. Something more substantial soon follows: three developments of method that are of considerable significance to any demonstration of White's seriousness of purpose. They concern the differentiation of states of flowering, the inclusion of a Latin alongside a common name, and the specification of a place of discovery. Each of these merits separate discussion, and I start with the last.

Record of place is first noted on 30 August with the supposition that the 'great purple snap-dragon' which White 'found in a lane at Empshot' had been 'thrown-out from some Garden', and is followed during the first half of September by recording plants on 'the Lythe', 'the steep chalky end of Whetham-hill', 'the bogs of Beans-pond in Wullmere forest', and 'in a farm yard at Faringdon'. Then, during a visit to his aunt (Mrs Snooke) at Ringmer in East Sussex, he noted plants 'on the downs', in 'a lane towards the sea near a village call'd Whiting', and 'in the pasture-fields at Ringmer', as well as observing an 'abundance of sea-plants on ye shore which [he] had not time to examine'. On his return to Selborne, White gave further attention to the sites that he inspected (for example, he notes plants 'in my beechen Hanger', 'in the hollow lanes', 'in Selborne-wood', 'in Mrs: Etty's garden', and so on); and he sustains a similar level of record during a trip to Oxford, which yields, in particular, a list of plants 'on the banks of the Thames as [he] walked from Streatly to Wallingford', and another obtained during a visit 'at the Physic garden'.

At first one is struck by the variety of these locations, but then, perhaps, mildly irritated at the inconsistent level of detail.[18] Yet, to expect from White an absolute consistency of detail — except perhaps in the later meteorological records of the *Journal* — is to misunderstand the particular form his English genius most typically expresses. Certainly, as with the plant locations under discussion, records will be made, but the methods and purposes associated with those records will probably satisfy, as here, a number of interests. For example, only on rare occasions does White ever deliberately investigate a natural phenomenon with a clear intention declared in advance, and with the necessary resources to hand. More customarily, he accepts what the natural scene offers as he proceeds in his daily round, and this is what happens in his botanising, as the records show. In some instances the *Kalendar* note appears to have been made in such detail that the intention is clearly an *aide-mémoire* to a site, as occurs on 31 October:

> Discovered the Ivy-leaved Sowthistle, or wild lettuce (Lactuca sylvestris Murorum flore luteo) in a most shady part of the hollow lane under the cove of the rock as you first enter the lane . . . on the right hand before you come to the nine-acre-lane . . .

On other occasions the content of the entry appears to focus less on site, although this may be mentioned, and more on the habitat and distribution of a plant. Compare for example a note for 17/18 September ('On the poorer parts of ye Sussex-downs I saw the smaller Burnet in plenty') with one dated 20 September ('Hawkweeds all ye Country over from the highest downs to ye lowest pasture-field'). Here, in both notes, what is being recorded is not simply a botanical fact, but White's learning 'on the ground' so-to-speak that the study of plants was more than 'a pursuit that amuses the fancy and exercises the memory, without improving the mind or advancing any real knowledge'.[19] As will be shown when we consider what White has to say in *Selborne* about the value of botanical study, to dwell constantly on lists of species and locations is to pursue 'a mere systematic classification' at the expense of studying plants 'philosophically'; and the first evidences for this attitude are embedded in these early notes.

Let us now turn to the language used by White in making these notes of late summer and autumn 1765; we shall discover a progressive mastery of standard terminology, analogous in some respects to the development noted above. At first (22 August) ordinary English terms are used, but at 30 August some Latin terms are introduced, and a two-term, partial, identification ('Ranunculus flammeus' for 'small spear-worth') appears.[20] These initial hesitant steps into technicalities[21] are followed in mid-September by the gradual introduction of very precise English descriptive terms (for example, 17–18 September: a rose with 'very beautiful small pinnated leaves' is recorded, as is 'a thistle with an echinated head, & little down[22] to ye seeds'). Then, during and after the 15–26 October visit to Oxford, plants begin to be noted in formal Latin—for example, 16 October: 'less stitchwort (Caryophyllus holosteus arvensis glaber flore minore)', 'the hawkweed called Hieracium echioides capitulis cardui benedicti; al: lang de beuf'; 2 November: 'dwarf-hawkweed with sinuated very narrow leaves (Hierachium parvum in arenosis nascens, seminum pappis densicus radiatis)'; 18 November: 'Discovered the common Spurrey (alsine, spergula dicta major) in pod, & bloom in a ploughed field: most exactly described by Ray'.

What these entries show is an increasing willingness, based on a growing competence, to record plants in the elaborate, descriptive Latin of John Ray. But Ray was not the sole authority White used; mention is made of John Hill and, prior to the visit to Oxford, some plants are noted in descriptions taken not from a flora but from Philip Miller's *Dictionary*. That White made reference to several texts and in later records employed Linnaean terminology indicates the width of his interests as well as his resilience. More importantly perhaps, it suggests the sound quality of the advice he received and his own good judgement. For an amateur botanist today to turn to several texts to begin an induction into the details of classification and nomenclature would be unremarkable, but in the eighteenth century authorities, although still quite numerous, were very much more restricted, particularly if native texts were sought. Indeed, almost up to the very time White himself began to study plants, 'the pocket companion of every *English* botanist' was Ray's *Synopsis*—in an edition prepared in 1724 by Dillenius.[23] This was progressively superseded from 1762 onwards with the publication that year of William Hudson's *Flora Anglica*. The volume had been prompted in part by Stillingfleet, who had quickly seen the power of the new Linnaean classification and the evident usefulness of Linnaeus's introduction of binomial nomenclature, and the resulting flora (the first British flora prepared wholly on Linnaean principles) was well received. White obtained a copy in 1765, marking in it eventually 439 species of his own finding.[24] Of these, 24 were later included in *Selborne* and accompanied by the remark that:

> To enumerate all the plants that have been discovered within our limits would be a needless work; but a short list of the more rare, and the spots where they are to be found, may be neither unacceptable nor unentertaining . . .[25]

Despite the hesitancy of this announcement (which, in fact, was not to be penned for a dozen or so years, and only after deeply-engrossing studies associated with a territory quite different from Selborne), White's devotion to botanical study in the middle 1760s was firm. However, of far greater interest than the mastery of terminology and the discovery of locations (even if they hinted at intriguing patterns of distribution) were to be reflections prompted by observations linked to the calendar.

As indicated earlier, the calendar White pursued was the natural calendar of vegetation, and not a calendar associated with the heavens. Astronomical calendars had been much favoured by classical authorities both Greek and Roman, but Stillingfleet had argued firmly in the *Tracts* that the relevance to English husbandry, subject as it was to the vagaries of a temperate climate, of a calendar based on the regularities of heavenly bodies must be doubtful.[26] Evidently White concurred. But to establish the new kind of calendar, that would reliably show the proper times for sowing, planting and harvesting, required the preparation of very detailed records—especially of the flowering of plants, the stage that was taken as the most critical in a plant cycle.

This being so, after 9 August 1765 White began to include this kind of information in his records. As with his progress towards other aspects of keeping a natural calendar, the way forward is initially hesitant but soon becomes confident and assured. For example, on 14 August White made a list of eight herbs 'in high bloom', and followed this two days later with:

> The large Aster with yellow thrums . . . begins to flower . . .
> The uncommon Aster with a black thrum blowing.
> The variegated Epilobium in bloom.

On 22 August note is made that seven wild plants are 'in flower'; then two very significant developments occur. First, several stages of flowering, both within a single species and in different species, are noted; second, the record begins to include the decay of leaves on trees. Selected entries from the *Kalendar* concerning both these stages of observation are:

27 Aug: Mich: daiseys begin to blow. Earth-nuts, & blue Devil's bit in bloom. Althaea frutex in high bloom: Ladies bed-straw just out.

20 Sept: Discovered plenty of the prickly rest-harrow (Anonis) & dier's broom, both in bloom & pod . . .

30 Sept: The beeches begin to be tinged with yellow.

1 Oct: The ashes, & maples in some places look yellow.

Following this, at the end of October (28th) a list of over 40 species 'still in bloom' is preceded by these remarks:

> A very smart frost that made the ground crisp, & has stripp'd the mulberry-tree & some ashes. The Hanger looks very much faded; & the leaves begin to fall. In general the new-sown wheat comes-up well.

Taken as a whole, rather like the citations given earlier concerning bird behaviour, these records denote a new focus on the natural scene. Whereas in the early years of the *Kalendar* there are almost no entries at all about flowers (except in association with a special venture, e.g. cockscombs; or with particular produce, e.g. growing of melons), the concern to observe and record stages of growth as a contribution to establishing a natural calendar becomes absolutely dominant. As shown above (28 October citation), the link with husbandry is made explicit; and the year ends with an announcement of the new cycle of natural growth, with the entry at 25 November:

> Discovered on a bank at Faringdon Filex elegans, Adianto nigro accedens, segmentis rotundioribus; a beautiful fern about six inches high: Pilewort, or ye less Celandine (Chelidonium minus) in it's first leaves; it blows in March, & April: The greater Celandine in it's first leaves (Chelidonium majus vulgare) & chervil in it's first leaves (Cicutaria vulgaris; sive Myrrhis sylvestris seminibus laevibus) called also wild Cicely, & cow-weed.

Specification of place of discovery, formal Latin alongside common English names, and state of growth, are here all present in the single paragraph. A calendar could now be tackled for a whole year, with confidence and conviction in the worth of the task.

The *Flora* for 1766, written in White's customary fluent and elegant hand, covers 61 small octavo pages, and is a delight to read. Over the year as a whole there are dated entries on more than 170 days, three-quarters of this total occurring in the six months of March to August. For a single day, entries range from the very brief (for example, 10 November: 'Lychnidea blows still') to the very extensive (29 April, 16 May, 22 July, where, for each date, over a dozen observations extend in each case to a whole page and more). The entry for 25 June is fairly typical, and one in which some of the important characteristics of the *Flora* can be seen (see Plate 4). Part of it reads:

> 25 June: London-pride, Geum folio subrotundo majori, pistillo floris rubro, blows.

> Common nettle, Urtica racemifera major perennis, blows.
> Moss-Provence rose blows.
> Narrow-leafed bulbous Iris flowers, Xiphion.
> Garden-Larkspur, Delphinium, blows.
> The horned beetle, called by the French Cerf volant, appears:
>   scarabs: max: platycerus, taurus nonnullis.
> Crow-foot Geranium, Geranium Batrachoides flore
>   caeruleo, in bloom. The blossom is very large, & shewy.[27]

The first things to note concern the flowers: from a botanical point of view there is an apparently indiscriminate recording of plants from the garden with those from the wild. Indeed, it might be argued that the entire entry has nothing to do with the flora of Selborne but everything to do with the flora of White's own garden at Wakes. Yet, as claimed above, the entry is fairly typical of many of the *Flora*; and we need, therefore, to adjust our expectations to White's intentions. Later, when he is offering advice to his brother John concerning the Gibraltar flora, he will sharply differentiate between what properly belongs to a flora and to a hortus (the wild and the cultivated), but at this stage of his investigation into natural productions he is interested primarily in pursuing the model advanced in Stillingfleet's *Tracts*— that is, the preparation of a calendar that would eventually be of value as a guide to rural and horticultural practice. This being so, what is most important is the recording of stages of growth. To mark these stages Stillingfleet had adopted a mode of literal recording (see his extract for 21 June given earlier), but White did not follow this example. Instead, he developed a system of verbal description. The range of terms used is wide: plants can spring, sprout, shoot, emerge, peep-out; trees leaf, bud, flower, cast leaves, become naked.[28] The keenest discrimination is reserved for stages of flowering. For the botanist, the plantsman and the apothecary, this is the most crucial stage of a plant's cycle, and in acknowledgement White devised a set of phrases that divided flowering into five stages:[29] 'buds for bloom'; 'begins to blow'; 'blows'; 'in high bloom'; 'going out of bloom'. Each of the phrases is subject to stylistic variation (for example, 'flowers' occurs as well as 'blows'; 'in full blow' as well as 'in high bloom'), but for entry after entry, plant after plant, the *Flora* records not merely the existence of a certain species, but the stage at which it had been observed, linked, of course, to a calendar date.[30] Taken in its totality, the strength of White's interest in his task and the meticulous way in which he attended to it cannot be questioned.

The plant record contains more than a listing of species for White's private purposes. After each native term occurs a descriptive name, in

Latin; and invariably in that order. This is in contrast to procedures adopted in a formal flora, where—as White knew from familiarity with scientific floras—priority was given to a universally accessible Latin name, rather than to names that might be only of regional relevance or local coinage. The document White was preparing is not, however, in spite of its title, a flora as such; it is a calendar, and one expected to be of relevance for his own area. To use local names was therefore no hindrance, and they might eventually be an advantage in terms of communication, especially if other informants were to assist in data collection. After all, a scientist needs to grapple with the particular as well as with the general, and dual naming recognises this.[31] Further, since it was a procedure sanctioned by the calendars in Stillingfleet's *Tracts*, there must have seemed little merit in changing the pattern.

But White was no slavish imitator. His *Flora* is not modelled exactly on Stillingfleet's own calendar; it is not modelled exactly on that of Berger, but draws from each as his own inclination and knowledge direct. This becomes obvious as soon as we move beyond the botanical record and consider what White was prompted to include in addition. Although the mention of flowers is paramount, accompanying the 700 or so common English names that occur for plants there is sufficient evidence, as the extended sub-title indicates, for the document in its entirety to be treated as more than 'A Calendar of Flora'. As one might expect, of these other entries most (about 100) concern birds, but there are well over fifty that indicate weather conditions and about the same number that focus on insect life. In this last category occurs the note at 25 June concerning the 'horned beetle' (stag beetle). As with plants, White's major reference texts for creatures are works by John Ray, and the inclusion here of a French term is therefore unusual. Yet, although such supplementary information is not common, it illuminates the author's personality. What happens with the stag beetle derives, I think, from White's interest in hunting. To discover that the French, whose prolixity[32] (and perhaps also their fancifulness?) White deplored, could conceive of deer that could fly ('Cerf volant' literally translated being 'the flying stag') must have touched his sense of humour—and in it went to the calendar.

Other creatures in the record include ants, snakes, frogs, fish, several species of fly, bees, bats, and butterflies. Alongside these are the birds, especially the swallow, house martin and rook. At first glance there is nothing strange in such an assemblage of creatures, but why, one begins to wonder, are finches almost absent from the record? Why do robin and blackbird merit no more than a single entry each? Where are the moles and the mice, the sheep and oxen, the cats, dogs and horses, and all the other common creatures of farmstead and house?

The answer is surely linked to the fact that the plant record (strictly understood) excludes exotics. True, there is a note on 4 November about pineapples,[33] but the clear and incontrovertible focus of interst is not on how such a species might contribute to a natural calendar, but the very reverse: the special conditions (in this case the necessity for heating the stoves) required to sustain the plants in the English climate. Such activities provide no guide for a Hampshire husbandman. A local economy responds to local conditions; the plants of the parish share a climate with the very food crops the whole project was designed to increase; it is, then, those plants that figure most prominently in the calendar.

If exotic plants were unhelpful to White's purpose, with creatures the focus could be different: there are more categories to consider. In face of climatic cycles, plants either live or die, those that live seeming for the most part to share patterns of growth—bursting into activity in the spring, and flowering and fruiting in summer and autumn. In the animal kingdom other options are available. Certainly, some creatures survive here and are active the year round, but others hibernate, and still others migrate. If the purpose of the calendar is to note the harbingers of spring, or the onset of winter, species that hibernate or migrate would seem to have more claim on our attention than those whose daily presence announces their adaptability. What we must conclude is that each time the *Flora* records a creature there is probably some specific reason for the observation. Consider again the stag beetle of 25 June. For English conditions, the size is unusual, and so is the mode of flight. But it is not these characteristics that prompt White to make the record he does. Rather, he will have been responding to the suddenness of their arrival, and the brevity of their stay.[34] Likewise, we can discover similar levels of comment in most of the bird entries.

The inclusion of birds as a class in the calendar can be attributed to Stillingfleet. They are mentioned by Berger, but with very little attention compared with that given by his translator, who felt that birds might be especially important. The strength of this attention is shown most clearly in Stillingfleet's preface, where he gives an account of the role carried by birds in classical culture, particularly in the field of augury.[35] Moreover, from biblical times birds (if one remembers merely the raven and dove of the Flood) had been viewed as possessing special qualities as communicators of conditions elsewhere than those obtaining for creatures rooted to the land, whether metaphorically (man himself) or literally (the flora of the earth). This being so, it is to be expected that White should include in his calendar many references to migrating birds, and on occasion associate them in the record with other 'co-incidences' (see Plate 5).[36] Careful scrutiny of the *Flora*, however, suggests that he may have found as much interest in other kinds of phenomena. Certainly, birds are part of the record, and importantly so, but the range of smaller creatures that find a place in the calendar is striking;[37] and a strong case can be argued that the task White set himself (after reading Stillingfleet) had re-awakened the lively interest he had discovered for insect life at the time of the shower of gossamer in 1741.

In this sense, it is worth concluding consideration of the *Flora* by noting the remarkable continuity of White's records both in time and place. The scenes of his childhood, and phenomena he had known from the time he had first walked the fields and lanes of his village, were never dulled by intimacy. Now, as a man in his mid-forties, he was content to record day after day, month after month, species of plant and of creature that had become almost as familiar as the scenery itself, whether in the daily round of his own parish, or during his several journeys — to his aunt at Ringmer, to relatives in London, or to his brother living on the edge of Salisbury Plain. Many of the species are recorded more than once: to catch, for birds, dates they first appear, when they first sing, when the young are fledged and when they first retire or are silent; and for plants, the stages of growth and flowering discussed earlier. Yet — and this is one of the most extraordinary aspects of what White achieves — whatever it is that the record carries it does so with a freshness of phrase, an edge of enthusiasm, that conveys a continual delight in the natural world.[38]

Admittedly, the *Flora* itself was a new kind of document for White. As with many a new venture, the breaking fresh ground might of itself constitute its chief interest. If this is so, however, it is not the whole explanation; and it was not at the expense of existing records. In 1766 the *Garden-Kalendar* continued, as it had done for the previous 15 years; and it provides a fitting point of departure for renewed attention to White's life.

CHAPTER EIGHT

On Equal Terms: 1766–1769

The effects that flowed from keeping the *Flora* for the whole year of 1766 were considerable. They became visible in the following two or three years, but one effect was concurrent and can be treated straight-away: the changed status of the *Kalendar*.

To attend, in the manner already discussed, to the amassing of data in connection with the natural calendar inevitably led to a division of records. With so much to enter in the *Flora*—flowers (wild, and cultivated; and including those on fruiting trees), insects, birds, the sprouting and decaying of foliage—there would seem little left to include in the *Kalendar*. White thought otherwise, and gave it his close attention for much of 1766. Certainly, there are changes: it is much shorter in overall length (by about a half or more) than his kalendars completed earlier in the decade, and the number of entries in the last three months of the year is insignificant, there being only six in all. Yet, from January to June, data are recorded on average every other day; moreover, the character and purpose of many of these data mark a considered change of viewpoint from that of earlier years.

Most of this material relates to two fields only: garden operations, and the behaviour of the weather. The former are unexceptional: they note preparation of the beds for cucumbers and melons (and give details of their culture), the sowing of beans and peas, and the planting of potatoes, globe artichokes and new fruit trees; and other routine activities reported in earlier years are also included. What is different is the weather record: not that such notes are a fresh departure (they occur throughout the *Kalendar*), and not because they suddenly incorporate numerical data (that is a development that does not consistently appear until 1768), but because they give to the weather an attention that is quite new. Partly, this is a matter of focus: in the absence of so many other kinds of entry, those that remain (records of planting and sowing, and of the weather) become more prominent. But that is not the whole truth. In earlier years the typical weather entry accompanied other kinds of data and rarely seemed to be recorded for its own sake; now something different happens. It is best understood by following a sequence of complete entries over several days:

25 Mar: Snow in the night, & Ice.

26 Mar: Rain in the morning from the S: 'till twelve; then the wind turned N: & there came a violent snow for six hours, which lies very deep on the ground; & is but a bad sight so late in the Year.

The wall-trees have been boarded, & matted all day; & the hot-beds have scarce been opened at all.

27 Mar: A very heavy snow all day; which by night lay a vast thickness on the Ground; in many places three feet. All the shrubbs were weighed flat to ye earth. The hot-bed was never uncovered all day; but the plants lived in darkness. The boards & mats were kept before ye wall-trees.

28 Mar: The snow melted in part with a strong sunshine: but it is still as deep as an horse's belly in many places.

The Cucr: plants look very well to day.

29 Mar: Warm air, & a swift thaw: yet ye snow is very deep in some places: all along ye N: field it is deeper than an Horse's belly.

Stopped-down the Succades: they are fine plants.

30 Mar: Snow goes away with a gentle rain.

1/2 Apr: Great rain.

Female bloom of a Cucumber blows-out.

3 Apr: Black moist weather: the Hot-beds want sun.

What is significant about these entries is not so much the prominence of the weather record, but its sustained presence. From the fall of snow on 25 March until the rains of the next month there is an almost daily record that gives a complete meteorological narrative, supported in many instances by the factual details of direction, duration, and depth of snow. Further, within that narrative the weather itself is both plot and hero—all else is subservient: the flowering of the wall-fruit, the 'bloom of a Cucumber', the very 'Hot-beds' themselves, attend the violence of the snow, the loss of the sun.

As a narrative whole these *Kalendar* entries are not untypical; there are several more that give a similar effect. Taken altogether they point to a firm conclusion, namely, that the achievement of the other documentary record of the year 1766, the *Flora*, should not be seen in isolation. Although physically it appears a separate document, in fact the decision to collect data designed to illustrate the natural calendar brought with it an enhanced awareness of the importance of meteorological concerns, as well as the incentive to record weather narratives in a more comprehensive fashion than had been accomplished formerly.[1] No longer is the weather seen as something incidental to natural processes (however drastic or favourable its effects might be), but as something that merits close enquiry—for, perhaps, both the rewards of studying it as a natural phenomenon in its own right, and the possibility of discerning ways in which its influence (whether joyous or baleful) could be foretold.[2] Confirmation of this new enquiry, and a detailed analysis of White's meteorological records, will form part of our consideration of the *Journal* (Chapter 9).

The effects that flowed from the 1766 *Flora* were not confined to meteorology. Much more importantly, they were to do with an attitude of mind. The demands of sustaining the *Flora* record had been substantial, but so was the outcome—especially, as will be shown, in confidence.

Secure in the occasional duties of his curacy at Faringdon, and exhilarated perhaps by the satisfactions obtained from completion of the 1766 *Flora*, White began 1767 by keeping only one record: the *Kalendar*. In some ways it is an oddly attenuated document; and—particularly since I have just commented on a growth of confidence—there seems a strange loss of focus, an absence of coherence even, as if White was no longer sure what kind of record he was keeping. Evidence for this occurs throughout the year: some entries are of the briefest, others are sustained paragraphs; notes concerned with the natural calendar occur alongside those that give accounts of garden activities; details of procedures adopted in brewing and wine-making[3] intermingle with report of the weather; descriptive remarks on a botanical expedition abut speculation on the relationship of the area of an earwig's wings (when adult as well as when immature) to its body weight; and the nest-building efforts of the nuthatch give way to lively accounts of harvest mice. Of all the annual records that find a place under the title *Garden-Kalendar* the record for this one year stands alone, uncertain of both function and audience. For such a document to come from a man imbued with a strong sense of eighteenth-century proprieties, it must have seemed faintly disturbing, even alarming. An explanation and, in 1768, a resolution were, however, to hand.

The beginnings of the explanation are found in the 1767 *Kalendar* with the adjacent notes, 'Went to London' (18 April) and 'Returned to Selborne' (12 June), for it was during such visits that White first emerged socially from the scientific seclusion of Selborne.

As far as extant documents reveal, indeed, as White himself claimed, he had, up to the end of 1766, pursued natural knowledge almost unaided. Granted he had imbibed the accumulated wisdom that was the knowledge of the countryfolk he met daily (though much of that was based on superstition)[4] and there had been the enthusiasm and interest of his own family,[5] but, in the main, what he now knew had been achieved through his own endeavours. Of course, books had been available and he had made good use of them, both Miller's *Dictionary* and Stillingfleet's *Tracts* having yielded rich fruits; yet, for a man who felt that the test of a real naturalist was his skill in the field, something was still missing. What he most needed was to share his knowledge and test out his deductions against someone with wider experience, to find, as he put it, a companion who would 'quicken [his] Industry, & sharpen [his] attention'.[6] And that, exactly, was what happened.

After a brief hiatus (the first months of 1767), the momentum gained in completion of the *Flora* was followed over a space of a couple of years by developments of the greatest importance for White's progress in the field of natural history and for any assessment of how the country parson became author of the acclaimed *Selborne*. In chronological order, these developments (put quite baldly) were: beginning a correspondence with Thomas Pennant; receiving as a gift a copy of the *Naturalist's Journal*; visiting in London Joseph Banks and Daines Barrington, and welcoming to Selborne Richard Skinner, William Sheffield, and Mrs John White. Each merits separate consideration.

Thomas Pennant

The strong likelihood that White first met Pennant in London in April–May 1767 comes from his own pen, as does the probability that the meeting occurred at the bookshop of his brother in Fleet Street. Over the years Benjamin White had established a considerable reputation as a seller and publisher of books dealing with natural history; and it was quite common in the eighteenth century for a specialist shop to serve as a mecca for enthusiasts of the subject.[7] Pennant was often in London: already he was an established naturalist (inspired by a copy of Willughby's *Ornithology* when he was aged twelve years), writer (he had corresponded with Linnaeus since 1755), traveller (he had been on tour in France and the Low Countries in 1765), and author (his *British Zoology* had been published 1761–6); and he had been elected FRS on 26 February 1767. At first sight White would seem to have little to commend him to Pennant's attention. On the Continent the latter had met several of the leading scientists of the day, men as distinguished as Haller and Daubenton, Pallas, Brisson, and Buffon (with whom he had stayed), and in England he was acquainted with everyone who claimed an interest in natural knowledge. Nevertheless, the meeting was auspicious: partly because White was still excited by many aspects of the Linnaean programme (as advanced in Stillingfleet's *Tracts*), but primarily because of a methodological procedure adopted by Pennant himself in collecting information on natural history in particular localities. This took the form of circulating likely informants with a list of printed 'Queries', in which respondents were asked to furnish information about local topography, migrating birds, and other interesting natural phenomena.[8] Apparently, however, it was only after White had left London that Pennant realised what a rich seam of data there was in Selborne, ready, and even eager, to be mined: and so he pressed Benjamin White to urge his brother to correspond. Initially White must have been cautiously suspicious (his last communication with a national figure, the Lord Chancellor, over the living at Bradley, had ended uncompromisingly); but his pleasure at discovering that he had something to communicate soon outweighed any doubts, and on 10 August 1767 he wrote:

> Sir,
>   Nothing but the obliging notice You were so kind as to take of my trifling observations in the natural way, when I was in town in the spring, & your repeated mention of me in some late letters to my Brother, could have emboldened me to have entered into a Correspondence with You.[9]

Then, after expressing delight in the opportunity now granted him (of sharing his knowledge), the letter continues with discussion of natural phenomena known to White. Some of these phenomena incorporate aspects of permanent speculation on White's part, most notably whether swallows and related species, especially the young from late broods, remained in England throughout the winter in a torpid state. Other phenomena, however, deriving directly from his meeting Pennant, are matters of report: examples here are his sending Pennant a bird (a falcon) for identification and his apologising for having 'had no opportunity yet of procuring any more of those mice I mentioned to you'. These last were specimens of harvest mice, *Micromys minutus*.[10]

In this first letter to Pennant (so typical of all that follow), what is of far more importance than the content (in terms of species) is the way White addresses the several topics. His manner is revealing and in complete contrast to Pennant's own procedures. Right from the beginning White's enthusiasm for a descriptive mode is transparent. Also, he invariably shows that the source of what he describes is based not on book-learning (although reference to Linnaeus, to Derham, and to Pennant, indicate that in the previous months he had extended his studies beyond Stillingfleet and Hudson),[11] but on the evidence of his own eyes. Hence the accounts of various creatures are accompanied by as many factual data as are conveniently to hand: usually these take the form of external measurements (of weight, of length of tail or wingspan, for example), but later he realised the value of dissection and in some letters incorporated remarks on the contents of avian stomachs, on progress towards laying and on other internal anatomical features.[12]

Yet the tape measure, the scales and the knife were subsidiary to what White found most absorbing: the life and conversation (as he called it) of creatures. Almost every letter he penned is studded with a jewel or more of behavioural observation. Consider this from the 10 August letter:

> The Stoparola [flycatcher] of Ray (for which we have no name in these parts) is called, I saw, in yr Zoölogy the cobweb-bird. There is one circumstance characteristic of this bird, which Ray & Linnaeus take no notice of; & that is, it takes it's stand on the top of some stake or post, from whence it springs forth on its prey, catching a fly in the air, & hardly ever touching the ground, but returning still to the same stand for many times together.

And this from his fourth letter to Pennant, of 22 January 1768:

> We have in the winter vast flocks of the common linnets, more I think, than can be bred in any district. These, I observe, when the spring advances, congregate on some tree in the sunshine; & join all in a gentle sort of chirping, as if they were about to break-up their winter quarters, & betake themselves to their respective summer homes.[13]

When carried over into *Selborne*, it is vignettes of this kind that contribute vividly to White's success as an author: what is described is acknowledged as what happens to every observant reader—White's truth becomes the reader's truth also.[14]

I have chosen to quote two avian anecdotes since they are representative of White's enduring pleasure and accord also with Pennant's zoological interests. Having launched the correspondence, however, White made sure that his own most recent and substantial investigations, incorporated in the *Flora*, became part of it as well. The day after first writing to Pennant in August, he had joined with his brother Thomas on a botanical foray to Bean's Pond, a few miles east of Selborne; so successful was the trip that he re-opened the 1766 *Flora* to record in about half a page several plants discovered on the margins of the water.[15] It is then reassuring to read in his second letter to Pennant, of 9 September 1767, some of the most notable results of his botanical investigations. After remarking that the countryside he lived in was 'a very abrupt, uneven Country, full of hills, & woods; & therefore full of birds', he continued:

> Our great variety of soils makes it excell also in Botany. We have some curious plants, such as the Lathraea squammaria, monotropa hypo-pitys, vaccinium oxycoccos, vaccinium myrtillus, drosera rotundifolia, ophrys nidus avis, &c: & a very great variety of the common sorts of plants: not to mention numbers that must escape one pair of eyes.[16]

Important as this list is, including as it does a plant that was to provide access to a future president of the Royal Society and sometime circumnavigator of the world, it is the following paragraph that is of most interest for revealing White's approach to his correspondent:

> I should take it as a favour if you could procure me some seeds of any of the following plants, which we never see in the south: viz: the Cypri-pedium calceolus, pinguicula vulg: polygonum viviparum, thlaspi montanum, actaea spicata, rubus chamaemorus, or Andromeda polifolia . . .

Since, White seems to be saying, he was providing Pennant with information and specimens, then in return Pennant might search out what would be satisfying to him, White. Put thus, the relationship assumed by White is decidedly two-way, evincing even an equality of expectation. For a few years this may have been so, but as their relationship grew, the open-hearted and generous White was to discover that Pennant's reciprocity was but borrowed plumage and that he had scant respect for scientific priority.

In autumn 1767, however, such clouds were in the future, and the country parson was more concerned to demonstrate the integrity of his information and to respond to the challenge laid upon him by Pennant's interest in the entire fauna of the Selborne district. To meet this interest White had to venture into areas of natural history that previously he had taken little notice of, most notably fish. Not unexpectedly, an occasional error of identification creeps in. Later, substantive comment will be made on the authenticity of White's records (and of *Selborne* itself), but at this stage it is worth noting White's own concern for accuracy. Indeed, so important did he judge accuracy to be (in the pursuits he was now following) that early in his correspondence with Pennant he openly admitted to an error—in confusing one species of lamprey with another—and wrote:

> I desire to retract it [the identification] . . . Nothing would vex me more than to find I had misled by my inaccuracies a Gentleman who intends to favour the world with his researches into natural History.[17]

Problems of identification were not the only difficulty. Pennant's enthusiasm for the knowledge he was now receiving from this new correspondent had prompted a far more broadly-based investigation of the natural world at Selborne than White had resources to handle. Inevitably, therefore, the *Kalendar* for the second half of 1767 was used to carry the miscellaneous records referred to earlier; White had no other document available to him. As I have already hinted, however, a resolution of the difficulty was imminent and it provides the key for discussion of the second development crucial to White's public acknowledgement of the world's growing interest in natural history.

Barrington's *Journal* (1767)

The first information we have in White's correspondence about Barrington and his *Journal* occurs in a letter to Pennant early in 1768:

> Your friend Mr Barrington (to whom I am an entire stranger) has been so obliging as to make me a present of one of his Naturalist's Journals, which I hope to fill in the Course of the year.[18]

Evidently Pennant, who must have known of this recent publication, had passed White's name to the author, the Hon. Daines Barrington, and the gift was the outcome.

Barrington himself was a lawyer by profession, sometime Recorder at Bristol and later, 1778–85, justice at Chester. He had wide interests: in literary matters (he was a friend of Dr Johnson and Boswell, and corresponded with Bishop Percy), in antiquities (he held office as vice-president of the Society of Antiquaries), and in natural history and gardening (he wrote an historical account of the latter). He was also fascinated by aspects of sailing and geography and devoted a major portion of his non-professional publication, *Miscellanies* (1781), to discussion of the possibility of circumnavigation of the North Pole; he was admitted FRS in 1767. Such a range of enthusiasms was not unusual in the eighteenth century, but in Barrington's case it seems to have resulted in few permanent achievements — and no biography. This is a great pity, for if his encouragement of White is any guide it is in Barrington that we might find an exemplary figure capable of illuminating much that happened in several cultural fields in the later decades of the century. He did, however, design the *Journal* (copies of which White used yearly from 1768 until his death in 1793), and that in itself is a substantial memorial.[19]

That White welcomed the *Journal* and responded keenly to its demands is evident from the assiduousness with which he completed it; and that illustrates quite properly two sides of his nature. All the previous records he had kept had been something of a private indulgence, both the frequency of entry and the content lying at his own decision. But Barrington's *Journal* was different, and placed on him compelling demands: he was to meet them magnificently.

In planning the *Journal*, Barrington had shown that he was conversant with contemporary writing about the natural calendar: as he remarked in his Preface, 'the general utility' of what he had devised was sanctioned by 'the Calendar of Flora . . . amongst Mr. Stillingfleet's Miscellaneous Tracts'. For White to find that the *Journal* bore such a close relationship to a work that had already made a profound impression on him must have been an additional incentive, and he lost no time in beginning to complete it on a daily basis. The detail required was considerable. Each page is designed to contain observations made over one week, and is therefore ruled horizontally to provide seven sets of data; vertically, there are eleven columns, which fall into three groups. The first group comprises six columns that demand geographical, calendarian and meteorological information (including readings from a thermometer and a barometer, wind directions, and rainfall), the purpose of all the data being to provide a context for the interpretation of records in the next three columns, the second group. These are the heart of the *Journal* for they ask for observations about when certain phenomena first occur; more particularly, they require observation of 'Trees first in leaf', 'Plants first in flower', and when 'Birds and Insects first appear, or disappear' — data fundamental to Stillingfleet's intentions as announced in the *Tracts*. Lastly, there are two columns that afford an opportunity for further observations, in connection with 'fish, and other animals', and of a more miscellaneous nature, such as 'Celestial Phaenomena . . . Memorandums made with regard to seeds being first sown . . . with regard to blights, and other accidents [and] . . . notice of the planting of trees'.[20]

Remarkably, in spite of the extraordinary variety of demands made by the *Journal* almost every category of data required had on one occasion or another found a place in one of White's earlier journals. Now there was an authentic programme of records to be made within the confines of a single document, and all his own many interests could be brought together and, possibly, be of value to posterity. This public nature of the endeavour had been affirmed by Barrington:

> from many such journals kept in different parts of the kingdom, perhaps the very best and accurate materials for a General Natural History of Great Britain may in time be expected, as well as many profitable improvements and discoveries in agriculture.

> It is hoped that the use will not be only experienced by the Naturalist and Farmer, but by the Gardiner.[21]

This potential for the *Journal*, for amassing data relating to a particular locality, was not lost on White. Acknowledgement of its importance figures frequently in his letters, and it is clear that a predilection for first-hand evidence (something he might have inherited from his father's legal training) was an important element in promoting the future correspondence with Barrington. But at this stage such a development was more than a year away and he first tried to engage Pennant in a similar fashion. Writing early in 1768, he remarked:

> As in one of yr former letters You expressed the more satisfaction from my correspondence on account of my living in one of the most southerly counties: so now I may return the compliment, & hope to have my curiosity gratifyed with regard to the first article by yr living much more to the North.

The gratification White sought centred on winter flocks of chaffinches which, so he had discovered, were composed mostly of female birds; he hoped that Pennant would be able to tell him whether the case was the same in the north, for then 'one might be able to judge whether they come over from ye Continent, or whether our female flocks make their migration from the other end of the Island'.[22] Although Pennant was unable to help with this enquiry, he formed a favourable impression of both White and the richness of fauna available in White's locality and arranged to visit Selborne in the spring during a journey to Goodwood, near Chichester, to view some specimens of moose-deer that had recently arrived there from Canada.[23] Unfortunately, however, he had to cancel the journey. White's disappointment was strong; writing on 19 April, he remarked: 'As I had set my mind on the pleasure of yr conversation, so I was in proportion disappointed when I found that you could not come.'[24] In fact, so intensely did he feel the lost opportunity that he extended an invitation in the most persuasive way he knew: by adverting to the appeal of Selborne:

> But [he continued] as yr business may be over now, I shall still live in hopes of seeing you at this beautiful season, when every hedge & field abounds with matter of entertainment for the curious. If you could come-down at the end of this week, or the beginning of next, I should be ready to partake with you in a post chaise back to town on the second of May.

It was not to be. Neither Pennant nor Barrington (the two principal recipients of the correspondence underlying *Selborne*) was ever to visit the village, and White had to pursue the flora and fauna of his home territory almost alone. He could, however, correspond, and he could make visits, and, as his letter to Pennant proposed, he travelled to London early in May 1768.

It was to be a propitious visit. Although he went without a specimen of *Lathraea squammaria*, toothwort (promised for a friend of Pennant's, the young Joseph Banks), he did take his *Journal*—filled with four months of data—and met in person its designer, Daines Barrington. Furthermore, and this is of greater consequence than is usually claimed, he was to find that the horizons for his understanding of natural history were to be stretched beyond this island, and around the globe.

Visits and Visitors

It is not uncommon for biographers of White to stress the familial nature of White's visits and visitors, apart that is from the professional concerns centred on Oxford.[25] There is much sense in this view, particularly if one considers on the one hand the many occasions he visited brother Henry (at Tidworth, and Fyfield), brothers Benjamin and Thomas (in London), and his aunt Rebecca (at Ringmer), and on the other hand the evident delight he obtained from welcoming relatives (including a growing band of nieces and nephews) to the family home in Selborne. Yet, if we think not of the village of that name, but of the volume that is to be *Selborne*, a quite different kind of visit and visitor assumes an importance far in excess of any numerical quantification that may be made. This is especially so in the period 1768–9.

In his very first letter to Pennant (10 August 1767), White had plaintively observed:

> It has been my misfortune never to have had any neighbour whose studies have taken the same bent: so that . . . I have made but a slender progress in a kind of knowledge, which I have been attached to from a Child.[26]

That was now to change. Within the space of a few years he was able so to broaden his experience of what was happening in the forefront of contemporary studies of natural history that his own knowledge and skills assumed a perspective that was new and original.

This adjustment of White's awareness can be centred on the *L. squammaria*. On the occasion of his visit to London the previous year, he had called on Joseph Banks, recently returned from a voyage across the Atlantic to Newfoundland and Labrador and an enforced Christmas, during the return trip, in Lisbon. The gap in age, experience and demeanour was considerable, and, although White had taken with him some specimens for discussion, conversation was not easy.[27] Banks, however, was sufficiently interested to follow up the visit in order to pursue 'Ornithological Converse'.[28] Evidently, other matters were talked about (perhaps a brief journey Banks had made in the spring to Kent, and one he was shortly to make into the west country); and there must have been some discussion of plants. At any rate, early in the next year (March 1768) White wrote to Pennant:

> Pray give my humble respects to Mr: Banks & tell him I shall not forget him next month with regard to the Lathraea squammaria.[29]

Unfortunately (or so it must have seemed at the time), White was unable to fulfil his promise. The whole of March was an exceptionally dry month, and the opportunity of supplying the plant slipped past:

> Selborne, April 21, 1768.
>
> Sir,
>   Lest you should suspect that I forget my promise, I take the liberty to acquaint you that either the unusual dryness of last month, or some unknown cause, has retarded the blowing of the *Lathraea squammaria*; it does not yet appear above ground as usual. When it had appeared I should not have failed to have sent you a specimen in a pot, with the *Coleoptera* its constant attendants; but now I find last night, by a letter of Mr. Pennant, dated from Chester, that you are going to leave the kingdom again in pursuit of natural knowledge.[30]

By this same letter from Pennant (which he had mentioned to Banks), White learned of the cancellation of the intended visit to Goodwood to inspect the moose. So intense was his regret at this news, so keenly had he anticipated being able to welcome his new friends to Selborne, that he was unable to restrain mention of his loss in his letter to Banks, and continued:

> I was greatly in hopes once that both you gentlemen would have honoured me with your company this spring; but now it seems . . . I must plod on by myself, with few books and no soul to communicate my doubts or discoveries to.

To write thus, to a man whom he barely knew, reveals an aspect of White's character not customarily available in his correspondence, and one that should be dwelt on for a moment or two longer. By initiating the exchange of letters with Pennant, White had offered to a stranger judgement on what he most loved—the beauty and variety of the Selborne scene. For the correspondence itself to have been reciprocated must have been satisfaction enough; but then, to discover that this acclaimed luminary intended to bring a friend to Sussex and, so White hoped, planned to visit Selborne itself (exclusively, we need to remember, on the evidence of White's report) must have been—almost—joy unalloyed. In consequence, when the cancellation reached him, it is no wonder that his remarks to Banks conveyed an undisguised regret.

Yet, White's faith in what his native village could offer was stronger than any personal chagrin, and he ended his letter with a description of what Banks was to miss:

> The district round this village is, I believe, a fine field for botany; we have great variety of aspects and soils, and many rocky lanes for capillaries; not to mention chalky hills, sands, bogs, clays, and vast woods. Here are moors that have hardly been trodden by the foot of a real botanist, where I should suspect some rare matters may be discovered. As to summer birds of passage, we have a good variety. I think I can shew 16 or 17 species round the village, among which are three species of the *Motacillae trochili*. The vast large bats are beginning to appear, some of which I shall endeavour to procure.

Whether the magic, if that is not too strong a word, wrought its spell on the recipient is unrecorded. We do, however, know that at the beginning of the next month White travelled to London, and again visited Banks. Writing later to Pennant, he remarked:

> Even Mr: Banks (notwithstanding he was so soon to leave the kingdom, & undertake his immense voyage) afforded me some hours of his conversation at his new house, where I met Dr: Solander.[31]

It is, perhaps, easy to labour the point, but White was remarkably affected by the acquaintance, and not least when, in the autumn, he saw a report in the press that Banks had finally set off to join with Captain Cook in the voyage of the *Endeavour* to the southern seas.[32] At this time he wrote to Pennant:

> When I reflect on the youth & affluence of this enterprizing Gent: I am filled with wonder to see how conspicuously the contemp[t] of dangers, & the love of excelling in his favourite studies stand forth in his character. . . . I cannot divest myself of some degree of sollicitude for his person. The circumnavigation of the globe is an undertaking that must shock the constitution of a person inured to a sea-faring life from his childhood: & how much more that of a land-man?[33]

Such solicitude, however, did not hide from White the possible merits of the expedition, at least in the field of natural history:

> May we not hope [he continued] that this strong Impulse, which urges

> forward this distinguished Naturalist to brave the intemperance of
> every Climate; may also lead him to the discovery of something highly
> beneficial to mankind? If he survives, with what delight shall we peruse
> his Journals, his Fauna, his Flora? . . . if he fails by the way, I shall
> revere his fortitude, & contempt of pleasures, & indulgencies: but shall
> always regret him, tho' my knowledge of his worth was of late date, &
> my acquaintance with him but slender.

Although his acquaintance had been slender, the effect on White was
strong. To have initiated correspondence with Pennant was one
thing; to have begun making entries in Barrington's *Journal* another;
for both demanded of him nothing more than the execution of skills he
had been developing over several decades. With Banks the demands
were on a totally different level: suddenly, he had to come-to terms
with a man prepared to risk all in the service of science. In view of
White's regard for the civilities and decorous rhythms of eighteenth-
century rural Selborne, that was a momentous challenge.

Evidence of the outcome takes two forms. Firstly, at periods
throughout the three years of the circumnavigation, White refers in
correspondence to the progress of the voyage as reported in the
newspapers; and on its successful completion he bade Pennant: 'Pray
present my humble respects to Mr: Banks, & tell him I heartily
congratulate him on his safe return from his astonishing voyage! The
world expects great Information from his discoveries during his
circumnavigation.'[34]

This is clear enough; the second outcome is less direct. There
appears to have been no renewal of correspondence between White
and Banks; indeed, there is no obvious evidence that they ever met
again—and yet, this absence of communication in the 1770s and
1780s should not be equated with an absence of influence. Attractive
as he found Banks, and admiringly as he wrote of the risks Banks was
prepared to undergo, the whole experience (for White) of meeting the
famous traveller acted as a kind of negative spur, and emphasised the
merits of his own modes of approach to nature. To tour the world
might lead to astonishing collections, but the real work of a faunist
must be to ascertain the life and conversation of animals: this was
something that required patience, careful observation, skills of identification and record, and the ability to translate what was seen into a
form, that is a language, that would communicate the behaviour
accurately. And White had these qualities in abundance.

Support for this approach to the natural world was readily to hand
in London, in the same May that White had again met Banks. With
the months of records he had already assembled in Barrington's
Journal secure on the page, he was in a confident position to be

introduced to the man who focused so precisely his own inclinations and approach:

> While in town [White wrote to Pennant, June 1768] I was often in company with yr friend Mr Barrington; & cannot say enough in commendation of the candor, & affability of that Gentleman.[35]

In their own fashion, these meetings in London with Barrington were historic occasions. By contrast with Pennant, Barrington himself thrilled to advising, to helping, to encouraging people—not for any personal acclaim, but in the interests of scholarly enquiry, civility, and a genuine curiosity about the world. It is in large measure to Barrington's 'affability' that we owe the most substantial, the most varied, indeed, the most engaging, of the three sets of letters that comprise *Selborne*.[36] But developments of this nature were in the future. For now, let us note the importance of these early meetings with Barrington, and resume our account of visits and visitors in the period 1768–9.

After White returned to Selborne in May 1768, there was much to attend to. As well as the garden to prepare for its summer beauty, letters to write, and clerical duty at Faringdon, there was more travelling to plan. In July he was to visit Fyfield, in September Chilgrove; and in October he was to be in Oxford for a week, having already been there earlier in the year in connection with the election of a new provost at Oriel.[37] In the midst of this, he was to urge Mulso and family to visit Selborne, receive company (including Dr Huddesford, President of Trinity College, Oxford),[38] and continue entries in the *Journal*. From the activities associated with this last he derived especial satisfaction, and although he declined an invitation to visit Pennant (in Flintshire) he must have been gratified to have Mulso remark early in 1769:

> I am not at all surprized at the Pleasure you take in your Pursuits of natural knowledge; very far from it; I know nothing more capable of satisfying the Curiosity of the human Mind, which is allways searching after Novelty, for the Subject is inexhaustible.[39]

Mulso of course was White's most enduring correspondent, but the visits White made, and the visitors he received at Selborne in 1769, far from curtailing the time he could spend on correspondence in fact contributed to a most productive year in his life for epistolary communication. What occurred is this.

Amidst the continuation of daily concerns, and family and professional visits (he was at Oxford in April and July, at Newton Valence for several weeks in late June and the beginning of July, and at

Ringmer in September), he made one visit (to London) and received two sets of visitors at Wakes that rival and, taken together, possibly outstrip in importance the events of the previous year. Each occasion initiated, or prompted, correspondence that was new and remarkably different from that begun with Pennant two years before in which White had predominantly been a reporter. Moreover, each of the series of letters that resulted is distinct in itself; that is, each series demonstrates considerable differences in role on White's part (in one he is primarily an enquirer, in another instructor), but whatever role he adopted he shows an enthusiasm and delight that become infectious.

Of all the series, the best-known is that deriving from the visit to London, during which he again met Barrington. As a consequence (and initially because they both shared a curiosity about birds), when White returned to Selborne and was at leisure, he wrote:

Selborne: July 6: 1769.

Dear Sir,

When I was in town in May I partly engaged that I would some time do myself the honour to write to you on the subject of natural history. And I am the more ready to fulfil my promise, because I see you are a Gentleman of great candor, & one that will make allowances; especially where the writer professes to be an *out door* Naturalist, one that takes his observations from the subject itself, & not from the writings of others.[40]

As a fragment of a complete letter, this is not unlike many similar paragraphs in letters to Pennant; and in that sense it is untypical. Over time, the Barrington correspondence developed an impetus that in spirit was quite different from that obtaining in the Pennant letters. These last are essentially programmatic. They set out to meet Pennant's requirements: a comprehensive account of the fauna in the Selborne area. In contrast, although the nature of the factualities in the letters to Barrington are unchanged, what emerges is a man sharing his most absorbing experiences and discussing what most intrigued him in the rich diversities of the natural scene. That this is so is shown in the correspondence being more long-lasting than that with Pennant: for White, to have Barrington as a correspondent was more satisfying, in all likelihood more companionable.

The other series of letters, three in all, derive from visitors coming to Selborne, but they differ in that the first two to be commented on consist exclusively of incoming letters, whereas the third is composed of White's own letters. On 10 August 1769 White noted in his *Journal*:

Mr: Sheffield of Worcester Coll: went into Wolmer-forest & procured me a green sand-piper, Tringa Aldrov: Tringa ochropus Lin: They were in pairs & had been seen about by many people on the streams, & banks of the ponds.

Sheffield was a fellow of his college (and a future provost) and was visiting Selborne with Richard Skinner, also from Oxford but from Corpus Christi College. In the light of White's regret at the cancellation of the projected visit by Pennant and Banks, the impact of these two visitors can scarcely be overestimated. His own appreciation is shown in a letter to Pennant shortly after:

Mr: Skinner . . . & Mr: Sheffield . . . have lately been with me for a fortnight: & are the only Naturalists that I have ever yet had the pleasure of seeing at my house. They are both excellent Botanists: & the latter makes a very rapid Progress in Entomology. There was great satisfaction in walking out with these men: because no bird, plant or insect came before them unascertain'd. One day we shot a Tringa ocrophus, which is a very rare bird in these parts.[41]

Partly as a consequence of this visit, White began to correspond with both these naturalists. Sadly, the only surviving letters are a few from Skinner and from Sheffield, White's own being no longer extant. If it is accepted, however, that there is likely to be a mutuality of detail in letters between mature correspondents, what survives is revealing. In both sets of Oxford letters,[42] Skinner and Sheffield certainly inform White, but they also treat him as an equal, interested in extending his knowledge, keen to solve a problem and conversant with the technicalities of the subject. Indeed, if one chose one word to characterise the letters it would be specificities: each letter is replete with exact detail and formal reference.

Somewhat similarly, a first glance at the final series of letters to be mentioned might also select out the level of technical detail as significant. But there is something else—of far greater influence. An entry in the 1769 *Journal* is again succinctly apposite:

28 Oct: Mrs: J: W: sailed.

The lady was Mrs John White, wife to White's brother John who was currently chaplain at Gibraltar. She had been in England visiting her parents, seeking a meeting with the Governor of Gibraltar (in connection with her husband's possible preferment), and bringing to Selborne a son (known familiarly as Gibraltar Jack) whose education White was to supervise. As far as White's studies in natural history are concerned there was a far more vital issue: she had brought from Gibraltar 'a box of curiosities', a small cargo of birds and insects. It was the outcome of White's having realised (on Pennant's prompting) the valuable service his brother could perform to natural history, and of his success in persuading John to address himself to a discipline that scarcely more than a year before White had confessed his brother had never previously studied.[43] From this first consignment (several more followed in the next two or more years), great things flowed. Of central importance is the correspondence.[44] All the extant letters are White's, but they show him in a new guise, prompting, cajoling, guiding, directing (as far as he could by letter) his brother's pursuit of the natural history of Gibraltar. Gone is the reporter of the Selborne fauna (that was his role with Pennant); absent are the wide-ranging, at times philosophic, speculations (of the letters to Barrington). In their stead is a continuance of the specificities associated with Sheffield and Skinner, but from a totally different perspective. With his Oxford colleagues White had been the student; now, writing to a younger brother, he was the teacher.

The confidence instilled by the accomplishment of the 1766 *Flora* had been burnished by the attentions of Pennant, Banks, Barrington, Skinner and Sheffield,[45] and was shining brightly. Now, instead of confining himself to the Selborne scene, White prepared, vicariously and at a distance, to launch into the charting of an unfamiliar terrain and climate. The mature investigator had been born.

CHAPTER NINE

The Naturalist's Journal

Preparation for dealing with the issues raised by the Gibraltar cargoes was not restricted to the confidence White gained from epistolary discussion of indigenous natural history. It also had to do with completion of the *Journal*; the format of this document has been discussed earlier and attention can now be given to the records themselves.

At a rough count the number of entries made by White over the years approaches 70,000 in all,[1] and comment must necessarily be strictly selective. The focus therefore of what follows will fall on the first of two phases of completion, namely on the records made in the period 1768–75. During these years White filled in the *Journal* very much in the spirit that Barrington intended, but subsequently (in the second phase, from 1776 onwards) his experience and interests led him to make certain modifications to the format; these modifications significantly changed the function of the *Journal* and will therefore be treated later. To restrict discussion to records made over eight years still, however, presents a very large number of entries — of the order of 2,500 annually. Analysis of this could take several forms. For example, comparisons might be made of the records of one year with those of another; meteorological data might be extracted and used to prepare daily weather maps;[2] or botanical entries might be assembled and used, in comparison with twentieth-century records, to determine changes in the Selborne flora over the past 200 years.[3] More important than any of these, at the present moment, must be enquiry into how White met, day by day, the demands made upon him by the printed, columnar notation offered in the *Journal*. As noted previously, all his earlier records had been made at his own whim; now, with the printed pages before him, he might have found the demand too exacting, or merely a matter of mechanical procedure.

To illuminate how he dealt with this problem, attention will be given to kindred groups of data over a period of time, say a year, so that conclusions about the characteristics of the record are grounded in sufficient detail to ensure that they are securely based. Discussion, therefore, will relate to the three groups described in the previous chapter: contextual data, data concerned with flora and fauna first appearing or disappearing, and miscellaneous observations.

Contextual Data

At the centre of these lie the meteorological records, but first there are things to note about the accompanying calendarian and geographical entries.

Of the data relative to the calendar, there is little to remark. Whereas in the early years White entered a year-date at the head of every page, he later dropped this practice and put it just at the beginning of each year's journal. As regards days and months, however, there is of necessity no such slackening of detail, and they are recorded meticulously throughout.

The geographic data are of more interest. Two entries are called for: information about 'Place', and a description of 'Soil'. Today, we know the chemical basis of the links between geology and the kind of habitat available in particular areas, and hence the way in which soil type contributes to and in some cases determines the flora and fauna of a region. In White's time this basis had still to be discovered, and although many experiments were conducted which related soil type to the success of certain agricultural and horticultural practices the focus was essentially descriptive.[4] This is shown clearly in White's records. Earlier we have seen the care he took to prepare beds at Wakes for various gardening operations, and in the *Journal* he included occasional observations such as that at 8 November 1771 where he noted the especial suitability of the 'brick-loam' at Ringmer for producing 'good apples, pears, & grapes'. But, so far as the requirements of the columnar notation invite descriptions of soil in connection with observation, he makes an infrequent record, noting for example in the 1768 *Journal* that the soil at Selborne is 'Clay' and at Fyfield 'Hasel loam on chalk'.

Of course, soil types are relatively constant and as with the year-dates there must have seemed little point in mere repetitive recording. With regard to place, however, a different view needs to be adopted. White's experience had already taught him that the natural phenomena of one region were not the same as at another, any more than the phenomena of one season were the same at another season. With this in mind, it is to be expected that the *Journal* would record place (and hence where observations were made, even if, as on a journey, retrospectively) with some care. Indeed, so carefully are the details entered that when White was travelling it is possible to determine not only the day of departure and arrival, but also—in association with other data—an approximate hour of the day.[5] In this sense, the *Journal*, as well as providing a vast record of natural observations, also performs the function of a diary. Although we may not know with whom he was staying or the purpose of a particular visit, what can be established from 1768 onwards is a detailed picture of the number of towns and villages that White visited, the frequency with which he stayed at this place or that, and thereby some idea of the opportunity he had to become familiar with a natural scene other than that at Selborne.[6]

White's reputation is securely based on his writings about a single region (Selborne), and it would be easy to assume that knowledge about phenomena at other locations was unimportant to him, even distracting. This might seem especially true in connection with meteorology. To discover the winter quarters of the swallow and martin that so brightened the summer at Selborne was of obvious interest, but of what possible use to an inquiry into the Selborne meteorology were weather records from elsewhere? Yet, as has been argued earlier in connection with his attitude to Banks's circumnavigation, knowledge of what occurred in other places acted as a powerful incentive to the closer scrutiny of his native region. Something similar occurred in relation to the weather. With the columnar demands of the *Journal* before him, his accumulated understanding (gathered over the years from newspapers, friends, and relatives—especially Thomas Barker in Rutland) of the varying climatic conditions within Britain acted strongly as a considerable spur to the collection of data for Selborne: and the record is there in the *Journal* for 25 years to prove it. So central did such data become to the whole endeavour of record-keeping that later he proposed submitting diaries of the weather at Selborne to the highly-respected journal *The Gentleman's Magazine*, for publication monthly.[7] Moreover, when he came to conclude *Selborne* he did so with seven letters (Barrington 60–66) that are devoted almost exclusively to discussion of aspects of the weather: the reward he obtained from such matters could scarcely be indicated more vitally.

What exactly, then, comprises the records he kept? What are their chief characteristics? The printed format of the *Journal* called for details in five columns, namely readings for temperature, and pressure, the direction of the wind, rainfall, and a verbal summary of 'Weather'.

One column, that for rainfall, snow, or size of hailstones, is quickly dealt with: it shows no record until May 1779, at which time figures in whole numbers (for example, '53'—representing hundredths of an inch) begin to appear; an instrument for collecting rain was kept on top of the church tower (see *Journal*, 31 October 1782). That White delayed so long before keeping a precipitation measure is surprising, especially since we know that his brother John kept one, supplied by Thomas Barker, at Gibraltar early in the 1770s.[8] Of course, to keep an instrument that will measure rain effectively is more demanding in time and site than to establish satisfactory figures for temperature and pressure (that a sufficiently open aspect was difficult to secure is indicated by an instrument being placed on the tower), and the answer may be that stronger persuasion was needed than the mere fact of a printed heading in the *Journal*. Confirmation of the value of rainfall figures came from reading Huxham, *Observationes de Aëra*,[9] which referred to coastal climates and devoted some discussion to matters White found of particular interest (see *Selborne*, Barrington 60). Moreover, Huxham, as well as what would be required by *The Gentleman's Magazine*, showed that his meteorological data, lacking rainfall figures, were deficient, and that a useful element of comparison with other climates was not at present possible. Some awareness that this was so appears in *Selborne* (Pennant 5) at a point where White gives a rainfall record for the village over seven years: there, he includes the observation that because his own 'experience in measuring the water is but of short date, [he is] not qualified to give the mean quantity'. Nevertheless, whatever the cause for the delay in collecting rain, the data show that he embraced this addition to his tasks willingly and reported at the end of each year a month-by-month summary to his brother-in-law.[10]

Of the remaining four meteorological columns, two—the readings of temperature and pressure—raise similar issues and are conveniently discussed together, but 'Wind' and 'Weather' will be considered independently of each other.

Entries in the columns for temperature and pressure derive from instrumental readings and are normally recorded numerically. Data from a thermometer are to be understood as a reading on Fahrenheit's scale, pressure being noted in inches and tenths of an inch.[11] White's usual practice seems to have been to record readings daily, although on some occasions he entered more than one reading and on some days there are no readings at all. Omission of readings at times when he was resident at Selborne is at first sight puzzling, since there appears to be no particular pattern to the omissions. From careful analysis, however, it is clear that in nearly all instances perfectly understandable explanations are available: these range from clerical duties in an adjacent parish and the annual Selborne fair (when he might have felt too busily occupied with other things to bother with his instruments) to something as simple as failure of one or other of the devices, or to a lack of time or opportunity occasioned by the day in question being one on which he began or concluded a journey. In sum total, the number of omissions is quite high: in 1770, for example, there are no numerical readings for either instrument on 72 days, which is an omission rate of nearly 20 per cent; and the instances lie scattered across the year.

Something similar occurs in relation to multiple readings, that is, days when White recorded more than a single reading for temperature and pressure. Again, if we consider the evidence of 1770, two numerical readings for temperature are entered on 38 days and for pressure on 20 days; further, on five days there are three readings for temperature. A pattern as broken as this needs explanation.

To establish definitive answers is not possible; but there are some clues in *Selborne*. During discussion of some 'unusually hot and dry' summers (Barrington 64),[12] White remarks that in Selborne's 'hilly and woody district, [he has] hardly ever seen [temperatures] exceed 80; nor does it often arrive at that pitch'. Many of the multiple readings for temperature occur in the summer (in fact, all the 1770 triple readings are in July and August), and it seems reasonable to conclude that on these occasions, in addition to making a routine daily record (usually in the early morning), White had an interest in catching examples of high temperature. To do this, he obviously had to take a reading at or around midday or early afternoon. If tentative credence is given to this argument, reassuring confirmation of its worth is found in an earlier *Selborne* letter (Barrington 63) during discussion of the severe frost of December 1784. In anticipation of a low night temperature 'we thought', writes White,

> it would be curious to attend to the motions of a thermometer: we therefore hung out two; one made by *Martin* and one by *Dolland*, which soon began to shew us what we were to expect; for, by ten o'clock, they fell to 21, and at eleven to 4, when we went to bed. . . . in the morning, the quicksilver of *Dolland's* glass was down to *half a degree below zero*; and that of *Martin's* which was absurdly graduated only to four degrees *above zero*, sunk quite into the brass guard of the ball; so that when the weather became most interesting this was useless.

There, I think, is the answer to all White's multiple entries for temperature and for pressure: they occur whenever something 'interesting' was expected. In the case of temperature this often meant extreme readings, at the boundaries of what was customary; for pressure something slightly different. Over time, pressure is considerably more stable than temperature, and to see why White made multiple entries can be assessed only in relation to the weather over several days. If this is done, it is apparent that he sought to record pressure when conditions were other than those expected for the time of year (which is quite different from what emerges from analysing the temperature entries, but equally 'interesting'), or when the weather was 'on the turn'. For example (taking data again from 1770), double entries appear on several days in January that carry the verbal note 'sweet': evidently, White was asking what occasioned such weather? What is the accompanying pressure? Other examples of double readings occur at times of 'vast' precipitation; and on 7 February, when a period of settled weather with temperatures in the 40s was succeeded by a 'violent' and 'vehement' wind, with frost and a temperature down to '29', double readings occur both on that day and the next. Similarly, a period of frost giving way to rain also leads to a double reading (9 March). From such examples, it must be seen that in addition to the normal single, daily reading of pressure (very often taken at midday) White added readings at times when the conditions were out of the ordinary.

What, then, can be concluded about White's instrumental records is that data which at first sight appear erratic and inconsistent in both the number and the frequency of their entries are a very human document. The keeping of the record possessed no absolute imperative: rather, there was an opportunity to keep as detailed and accurate a record, including records of the unusual and interesting, as circumstances—which comprised duties and responsibilities beyond those of a data-collector—permitted; a human, and not a mechanical, balance was what obtained.

Of the remaining two meteorological columns, 'Wind' and 'Weather', there is less to say. Both columns are exceptionally full: for example, in 1770 there are entries for 'Weather' on all but 16 days. Although we know there was a weather vane at Selborne (which White sometimes used: see *Kalendar*, 18 June 1764), the primary reason for the fullness of the record must be that data could be obtained without recourse to fixed instruments. This freedom from instrumental inspection meant that entries could be made elsewhere than at Selborne: and this is precisely what happened. At most places White visited, except for large urban centres such as Oxford and London, it is usual for the *Journal* to maintain records of both wind and weather. As for the detail of the entries, wind is noted in one of the eight principal points of the compass and weather in verbal, descriptive terms.

The number of entries on any given day is variable, because the record shows that White endeavoured to represent conditions throughout each day—and in the English climate they often change quite rapidly. This is best seen in connection with entries for weather, but could be shown also for wind data. On some occasions there is a single entry, such as 'fine day', but on others the descriptive terms provide a temporal sequence expressive of actual conditions. Good examples of this in 1770 are 'Sun. clouds. wind. clear. fog.' (for 1 April) and (on 10 October) 'Sun   hot   clouds   storm at night'.

Finally, it is worth noting the range of terms White uses. Casual inspection suggests that it is richly descriptive: the weather may be 'dripping', 'sprinkling', 'misling'; air can be 'soft', 'brisk', 'sharp', 'drying'; clouds 'vast', 'broken'; and wind 'hollow', 'fierce', 'tempestuous'. This last group is especially interesting. All the epithets attached to wind (and in the 1770 record twelve different terms are used) fall without exception into a class of terms characterised as vigorously positive, although the outcome of the described wind would generally be regarded as unwelcome. In contrast, if for the same year epithets applied to rain are considered, several scales appear to be at work: for example, within a scale of quantity, one might place 'vast' at one end, 'small' at the other; within a scale suggesting strength, 'driving' is opposed to 'soft'; and on a duration scale, 'steady' might be placed against 'spitting'. Of course, White was writing at a time when strict scientific vocabulary for weather conditions had scarcely been developed; indeed, if we think of clouds, the first thorough-going proposals were not published until ten years after White's death.[13] Nevertheless, White himself might have experimented with a more rigorous vocabulary, for the records themselves suggest that he developed a keen interest in meteorology. That this does not happen reminds us again of the nature of his personality, and of his attitude to his studies. A willed desire to originate material, or a determined insistence on the worth or comprehensiveness of his activity as an outdoor naturalist, would be alien. Far better, one can almost hear him say, to continue quietly recording what happened and to discover that of all the entries to be made about the weather the most frequent is 'sweet day'.[14] In such usage does White's own gentle warmth suffuse his writings.

Seasonal Indicators

In an earlier chapter it has been emphasised that the prime purpose Barrington had in mind when designing the format of the *Journal* was the collection of data in the expectation that it would reveal, to man's benefit, the natural calendar. The foremost non-contextual data for this purpose are found in the central four columns of the *Journal*. These comprise data in relation to the leafing of trees and appearance of fungi (column 7); flowering of plants, and growth of moss (column 8); appearance and disappearance of birds and insects (column 9); and 'Observations with regard to fish, and other animals' (column 10). This last is of some importance to White studies, and will be considered first.

There is, in fact, little mention of fish throughout the *Journal*. Early in the Pennant correspondence White was asked about the presence of such creatures in his neighbourhood, and he sent to Pennant's draughtsman in London 'a little earthen pot full of wet moss, & in it some sticklebacks male & female . . . some lamperns, some bull's heads'.[15] That White furnished such specimens must not be taken as an expression of secure knowledge. Shortly after this service he wrote that he wished 'to be better informed with regard to Icthyology', and continued:

> If fortune had settled me near the sea-side, or near some great river, my natural propensity would soon have urged me to have made myself acquainted with their productions: but as I have mostly lived in inland parts at a distance from great waters, my knowledge of fish extends little farther than to those common sorts, which our brook[s] & ponds produce.[16]

If the absence of frequent entries about fish is attributed to lack of opportunity, the same reason cannot explain the infrequent completion of the column for 'other animals'. Over the years these include a range of creatures—bats, vipers, hens, moles, and reptiles, for example; but the total in any one year is small (in 1770 only four; in 1771 only a single entry). In part, this is because observations of this kind become transferred to the miscellaneous column (column 11). Yet that is not the whole answer. Throughout White's work there is an almost total exclusion of any systematic, sustained account of what must surely have been the familiar: cats, dogs, horses, sheep, cattle, pigs, mice and other mammalian creatures of farm and field.[17] In itself that helpfully reminds us that the *Journal* is designed not for the agriculturist nor for the husbandman and gardener, but for the naturalist; after all, the volume is titled the *Naturalist's Journal*. This being so, it is somewhat disconcerting to find in the *Journal* as a whole many entries devoted to other farmyard creatures: cockerels, hens, buck rabbits, house pigeons, turkey cocks, geese, swan-geese, and peacocks are all examples in a single year (1770).[18] One explanation for this apparently contradictory procedure lies in the fact that the majority of these entries concern birds, which were always White's most absorbing study. There is, however, another focus, most clearly expressed in a letter to Barrington: 'The two great motives which regulate the proceedings of the brute creation are love and hunger'.[19] In illustration of this judgement, just as he sometimes took the knife to determine the diet of birds, so did he record behaviour that bore witness to activity 'necessary to the Conservation of the *Species*'.[20] That, in the main, such records (which are made mostly in the spring) are primarily confined to oviparous creatures need not detain us. To make similar comment regarding mammals, especially the larger ones, would have stretched the bounds of White's sense of propriety: instead, he chooses a proper silence.

One further aspect must be noted. Most of the observations concerning generative activity are made on domesticated creatures, rather than wild ones. Almost definitely this is not a deliberate policy, but an outcome of circumstance and attitude. Domestic fowl of all kinds were readily visible in yard and field, and observations could be made in the course of other tasks. To make the same observations in the wild would have necessitated a spirit of inquiry that was not to develop for many years, and it would be unreasonable to charge White with not engaging in it.

A similar division occurs in two of the remaining columns concerned with seasonal indication: trees and fungi (column 7), and plants (column 8). The number of entries in a typical year (say 1770 again) is not great: two fungi and no trees (although there had been about twenty in all in the first year of *Journal* completion, 1768), and in column 8 fifty species. As with some previous columns discussed, additional entries are made in the column for miscellaneous observations. This is especially true for trees: as well as recording notes about leafing, White began to enter the occasional observation about leaf-fall.

The plant record, small as it is, scarcely promises more interest. The most notable feature is the selectivity. Typically, there is no mention of flora resulting from a botanical excursion: instead, the flowering of fruit-trees is recorded as well as that of the cucumber (male and female separately); gorse as well as wheat; the green hellebore as well as the Michaelmas daisy. In fact, if we consider all the plant entries, it is difficult to detect any process that determined entry into the record other than chance observation and availability: the culinary and the ornamental, the cultivated, the naturalised and the wild, all are equally likely to appear, or not appear, whether or not (to recall the purpose of the *Journal*) they were thought to be especially prognostic. Indeed, if the *Journal* was the sole available source of data, it would be easy to hold the opinion that White lived in an age when differentiation of plants into the wild and the cultivated was regarded as unimportant. But that, in spite of authorities such as the notable Philip Miller advocating the transplanting of wild plants into the garden,[21] is obviously untrue. Fortunately we have on record a clearer indication of White's attitude. In *Selborne*, the main letter devoted to plants (Barrington 40) gives a listing of rare, wild plants, together with a locality within the parish. In addition, there is available in White's own words the very discrimination that the *Journal* record would seem to question. In a letter to Gibraltar, giving brother John advice about what botanical records to keep, White wrote:

> Plants cultivated in gardens must be carefully distinguished from spontaneous ones: because none but wild plants must appear in yr Flora; tho' an hortus [a listing of garden plants] at the end will be curious.[22]

Given that evidence, what must be concluded is that White was thoroughly aware of the interests of the botanist as distinct from those of the horticulturist, but that in practice—in the privacy of his garden and *Journal*—such distinctions were less tightly observed than they were when he addressed the world.

Lastly of the seasonal indicators there is column 9: 'Birds and Insects first appear, or disappear'. Inspection of the entries (51 in 1770, comprising 48 species: 23 birds, 9 butterflies and 16 other creatures) reveals above all White's ornithological skill. Without exception (and this is so in each of the early years of *Journal* completion) the bird entries exclusively record species that 'appear, or disappear' because of their migratory status. Moreover, the entries provide some of the most important data available about White's contribution to ornithological study. For example, the stock dove, *Columba oenas*, would be classed today as a widespread resident breeder, but from White's records it appears in his time to have been a local passage visitor at Selborne.[23] An aspect of status, dependent upon place of observation, is even more vital in connection with the ring ouzel, *Merula torquata* (= *Turdus torquatus*). In opposition to considerable scepticism on the part of Pennant, it was White who first established the bird's migratory status in Britain. More precisely, in the classification used today, he was the first to record that in parts of Britain (Selborne) the species was a spring and autumn passage visitor. On the basis of these data other observers were later to prove that for, say, Wales and Scotland the bird was a migrant breeder.[24]

Much of White's important work on birds was accomplished before the 1770s, so the vast majority of entries in the *Journal* should be seen as confirmatory of prior observation and knowledge. This applies not only in connection with status, but also to other avian behaviour: sequences of spring song, birds that sing as they fly, early breeding species, and so on. Such was White's certainty that he included lists of such behaviour in his very first letters of summer 1769 to Barrington (*Selborne*, Barrington 1 and 2). Interests of this kind (over and above details of arrival and departure) still find a place in the *Journal*, but are assigned to the miscellaneous observations of the final column.

Of the second and third groups, butterflies and other creatures, there is less to say. Scarcely any attention is given in *Selborne* to the former, but White was interested in 'noxious insects hurtful in the field, garden, and house' and wrote that a history of 'all the known and likely means of destroying them, would be allowed . . . to be a most useful and important work'.[25] The record shows many of the 'other creatures' to be noxious, and in the 1770 entries there are examples representative of 'field, garden, and house': namely, horse-bot (8 Aug.), rose-chafer (8 June), woodworm (7 Aug.). In addition, the 1770 listing includes bees, dragonflies, dung-beetle, daddy-long-legs and crickets (to the last of which *Selborne* was to devote three letters: Barrington 46–48), together with woodlice and millipedes.

'Miscellaneous Observations'

Lastly, as regards *Journal* columns, there are the entries under 'Miscellaneous Observations, and Memorandums' (column 11). It is a column White used regularly: there are frequent entries on very varied topics and often a note each day on more than one phenomenon. As has been remarked earlier, White sometimes used the column to elaborate or clarify information given in some entry made under a specific heading; but, in addition, there are numerous other entries. The topics can be grouped into four broad categories: birds, weather, garden, and other occurrences. The content of these first two might be thought to extend our knowledge of White's observations little beyond what has already been noticed, but this is not the case. As well as elaborations and clarifications, several new topics are included: for example, sequences of entry in connection with nest-building occur, as do remarks about winter flocks of hen chaffinches and attendant speculation about the absence of males.[26] In meteorology there are notes about many phenomena that have no place in the contextual structure (earthquakes, lightning, auroras, haloes, and blue mist, all figure in 1770),[27] and observations occur about the effect of the weather on plants, and on the husbandman and his activities—quite specifically in the field (or garden) as well as on travelling conditions:

'Paths are steady [firm]' in January; in February we learn that the 'Turn-pikes are dusty'; in March that the frost is sufficient to ensure 'Dirt bears horse & man'; and the rains in May mean the 'Cartway [the main street through the village] runs [with water]'. All the entries are of the greatest significance since they tell us about the context in which White carried out his studies.

This is seen especially in the remaining two categories, which relate to the garden, and to the agricultural year. Although the *Journal* is a naturalist's record, White (and I think Barrington) made no artificial distinction between what fell within the purview of the observer (anything of interest to the countryman that was concerned with natural phenomena) and what man discovered during his purposeful involvement in seasonal activities: both find a fitting place in the record. Just because he stopped making *Kalendar* records in 1767 does not mean that work in the garden had stopped, any more than it means that he had stopped brewing beer or overseeing the cutting and transportation of hay from his fields into the yard. Hence, the *Journal* maintains a considerable body of data about work in the garden: sowing, planting, staking, pruning, and all the other innumerable jobs of the gardener each finding a place when White felt moved to make the entry. Similarly, the agricultural year begins to show a securer face in the *Journal* than it did in earlier records, and entries are made that cover many of the routine operations: for example, ploughing, sowing barley, cutting sainfoin, hoeing turnips, harvesting hops.

Given this variety of entry in the last column and the sparse entries under columns for seasonal indicators, it is obvious that White found the *Journal* both valuable and, at the same time, restricting. The undoubted facility of record afforded by the contextual columns seems not to have been matched by a sustained completion that would contribute to man's discovery of the natural calendar—if that *could* be achieved by the kinds of record Barrington asked for. On the other hand, as the years passed White became more and more interested in making sustained comment about any aspects of the natural scene that caught his quick eye or attracted his discerning enquiry.

1774–1775

The date when White started to use the *Journal* for sustained comment cannot be determined precisely, although for reference purposes one might say it begins on 1 January 1776. From this date onwards, the columns of seasonal indicators are for the most part ignored; completion of data in relation to context is maintained, but then, written across the remaining columns occur entries that are often whole paragraphs of descriptive observation or speculative comment; briefer notes also continue.[28]

In one sense, these longer entries are not new: paragraphs of comment had formed the letters White had been writing about natural history ever since he had met Pennant. Even twenty or so years earlier in the *Kalendar*, there is evidence of a desire to record remarks in a more extended manner than is customary in a bare, factual diary;[29] and undoubtedly White had for many years enjoyed engaging in oral discussion with friends and chance acquaintances about the natural scene. What, therefore, needs explanation is not why White began to write longer notes in the *Journal*, but why, after eight years of patient recording within the columns of seasonal indicators, the decision was taken to cease making the record in that particular manner and to incorporate it selectively within more generalised comment.

An answer I believe is found in the *Journals* themselves, in the two extant *Journals* for 1774 and 1775. As noted previously, White had first met the designer of the *Journal*, Daines Barrington, in 1768 and had begun to correspond with him the following year; when occasion permitted and they were both in London, the clergyman and the judge arranged to discuss natural history. One topic of conversation must have been White's progress in completion of the *Journal*, for a known outcome was that White lent Barrington two of his *Journals* (those for 1774 and 1775) for perusal.

Barrington's response was enthusiastic; and he declared it forcefully on the actual pages of the journals by making numerous annotations. Whether or not White expected this reaction, he must have been startled to find on his meticulous pages not just an occasional prose comment, but also a liberal scattering of red crayon crosses and other markings. The typical comment (there are only eleven in all) is positively encouraging, and urged White to pursue some aspect of an entry further. For example, on a note about the first weighing of his aunt's tortoise at Ringmer (7 Aug. 1775), Barrington suggested the creature be weighed every year; and on a report (12–18 Feb. 1775) that a spoonbill had been shot recently in Norfolk, he enquired: 'From whom did you receive this account?' (see also Plate 6).[30] No doubt, this bias towards verification expressed the legal nature of Barrington's professional life and, on the few occasions that it occurred, was acceptable to White.

To the crosses and other markings, however, his response may have been very different. The frequency of marking is high. In the 1774 *Journal*—if the contextual data which attract no annotations are ignored—about one in every seven of the remaining entries is marked. This means that, when White came to discuss the *Journal* with Barrington, as assuredly he must have done, there were something like 70 items to reconsider; and on a wide variety of topics, for the crosses range over worms, insects, beer, truffles, tidal damage, birds, wells, and so on. That Barrington was so interested and so thorough in his attentiveness to the records would have pleased White greatly; and yet, beginning in January 1776, the character of the *Journal* began to change. There must be an answer to that!

Several factors appear to have been at work. The first relates to the intensity of Barrington's response to the *Journal* entries. To pass remarks on a few items in a year is one thing, but to find that clarification of one kind or another is sought on about 15 per cent of what one has written might have been more than White could bear. To prove that White responded in this way, however, is apart from the fact of the changed mode of *Journal* completion, difficult to document. The legal mind can be tenacious in its search for explanation and White's quieter, more responsive and pragmatic tendencies are considerably at variance with such an approach—as is shown throughout his records, and on into *Selborne* itself. Except perhaps for the *Flora* and a few fragments of private correspondence (none of which appears in *Selborne*), the tone of White's writing never achieves an awareness that some matters are commanded by an urgent, compelling and comprehensive necessity. The tone of his writing is in sharp contrast to what is conveyed in some of Barrington's own published writings and is one of the noticeable differences between the two men: if they had ever collaborated in a book (as Barrington suggested),[31] they would admirably have complemented each other's approach to the writing of natural history.

A second factor that contributed to White's changed practice in the *Journal* derived from his own experience. When he received the *Journal* and began the first year of its completion, the experience of the *Flora* must have moderated his enthusiasm with some moments of scepticism. As noticed earlier, there are several remarks in the *Flora* which suggest that he was fully aware of the many conditions—plant maturity, siting, temperature, rainfall—that affect the flowering of plants, and hence their appropriateness as indicators to prompt agricultural practices. No doubt, much of this knowledge derived from his experience as a gardener, but completion of the *Journal* columns devoted to such data from 1768 into the 1770s must have done little to disabuse him of what he may already have suspected: that his collection of data of this kind for *Selborne* would contribute little to the discovery of the natural calendar. And Barrington, so I judge, concurred: in his chief publication devoted to natural history (*Miscellanies*, 1781), he scarcely refers to the natural calendar, and spends no time at all on a consideration of the prognostic value of plants.[32]

Of course, White continued to make entries in the *Journal* about some aspects of nature that were linked with the seasons: in particular, he kept a keen eye on the arrival and departure of birds, noted which birds dust,[33] and recorded other matters that appeared to proceed in a regular succession—leaf-fall, frog migration and spawning,[34] for example—but such notes could equally well have been made under an enlarged 'Miscellaneous' column. The search for seasonal indicators, in the form envisaged, had failed; it was more interesting to engage in other kinds of inquiry.

That there were other kinds of inquiry to be pursued was not a wholly independent decision. It also accorded with developments of the time. Essentially, this concerned the influence of the two giants of natural history in the second half of the century: Linnaeus and Buffon. By the mid-1770s the impetus of the Linnaean desire to establish a bare description and ordered classification of the multifarious works of nature had begun to falter. In its stead had arrived a wish to establish an historical programme, which would enable men to understand how the world in which men lived (and all its flora and fauna) had come about. Quite how well White understood this approach—the leaders of the new study were primarily Continental rather than British or Swedish—is not revealed in his correspondence. The influence, however, is felt at several points in White's studies and had been noticed by him around 1775–7: 'it is', he wrote to a relative, 'the fashion now to despise Linnaeus'. This climate of opinion dictated how a book should be written and laid out: '. . . exchange *Classes* and *Ordines* [namely, Linnaean classifications and, one suspects, the concentration of Linnaean descriptions] for Chapters . . . throw . . . tables back into an appendix', White remarked.[35] In fact, these quotations are part of White's advice to his brother John concerning the latter's account of the natural history of Gibraltar. It is advice White followed himself by reducing, from 1776 onwards, the quantity of columnar data in his *Journals*, and by writing instead more sustained entries on a wider variety of topics. Further, it is advice he had been aware of ever since he had begun to write to his brother in 1769 about how to proceed in the study of natural history.[36] Hence, completion of what had been designed as the heart of the *Journal* (columns of seasonal indicators) was an aberration, albeit one from which White learned much.

The importance of these matters for the future *Selborne* is immense, and merits separate treatment. But before we leave the *Journal* there is one final matter, fundamental to all data collection, that needs discussion. How accurate are White's records? Can they be trusted? Do they, after two centuries, need interpretation or can they be accepted at face value?

Accuracy in the *Journal*

To ask questions about the accuracy of White's records may, to some readers, seem deeply offensive. Is not White, one can hear them say, noted as a naturalist, as a person seeking to understand more about the life and conversation of animals? What possible purpose could there be in his falsifying data? Further, was he not a clergyman living in a quiet country parish, seeking neither fame nor fortune? If the accuracy of this man's records is to be questioned, we may as well ask whether birds can fly or fish swim.

Quite so! Yet, there are questions to be asked: not perhaps about White's integrity, but about the understanding we must bring to his writing. In one sense, this is a matter of reliability: White wrote about a huge variety of things and it must therefore be perfectly proper to ask, say, whether his observations on plants are as trustworthy as those on birds, or whether the instrumental data about temperature and barometric pressure are as dependable as his prose notes that describe weather conditions. But, in another sense, what we are faced with is a problem of decorum, of convention even, in relation to the kind of writing we are reading—which, in turn, is a matter of understanding the nature of eighteenth-century literary and scientific expectations.

The relevance of this kind of awareness can be focused quite simply. Is *Selborne* the same kind of writing as the *Journal*? Do they share the same approach to the natural world? Can both documents be trusted to give the same kind of truth about things?

Notions of truth are notoriously difficult to discuss, but before we tackle the *Journals* (*Selborne* will be considered later: see Chapter 12) the attempt must be made. For our purpose, the key problem is one of language. What, we must ask, are the roles that language can play in the world? More particularly, how does language impinge on the world as it is observed and written about by natural historians? In response, we can say, speaking very broadly, that man uses language to carry out three distinctly different purposes. The first is concerned with prescriptive definition, with the creation of systems that are self-reflexive, internally coherent, conceptually translucent: a typical example here is a logical system such as pure mathematics. At the opposite extreme to this purpose lies poetry: language that is expressive of man's deepest emotions and which, in its most heightened form, can be mysteriously magical, hauntingly compelling. Between, so to speak, these two contrary modes and uses of language exists language used to get the business of the world done: that is, language employed to carry on our daily activity, exchange information, and pursue the richly varied activities associated with a multiplicity of interests and needs.

Clearly, the pursuits of a natural historian fall within this last purpose, language being used to classify and to describe events that occur in the world. Natural history events, however, are of many different kinds: from, to take examples in the *Journal*, the blossoming of a lime tree or the mating of swifts, on the one hand, to the brilliance of an aurora or the overflowing of a well, on the other. In other words, the natural historian of White's time, with interests as wide as his, had to record markedly disparate data, many of them requiring the closest scrutiny on the part of the observer. Events in the field of natural history (in the wild that is, not in the laboratory) are not random events; but neither are they open to any form of strict prediction. The arrival of a passage migrant on 11 April in one year (as swallows, for example, at Selborne in 1770) offers only limited guidance as to the likely date of arrival the following year, and that oak trees are stripped of leaves one June tells little about what may happen the next season.[37] Moreover, natural history is composed of more than biological processes: events of the kind mentioned occur within a physical environment, namely under certain weather conditions. We must not think that White's meteorological data are any different in status from his observations of flora and fauna. The weather is no more open to absolute prediction than is the arrival of a swallow, and that has to be remembered in making an assessment of the accuracy of White's records.[38]

To gain purchase on the level of accuracy deployed in the records of a natural historian two centuries ago is not easy. The best way would be to walk the lanes and paths of Selborne with White and to verify his observations as he made them, but that is clearly impossible. We must rely on the records he made, and our own knowledge. Comment will be made first on the content of the record.

At this level, attention must be given to the 'likelihood' of what White observed: that is, at the seasons mentioned and in the localities specified, is it 'likely' that one could have seen the birds White recorded seeing, the plants he said were in flower, and the many other living phenomena that have a place in the *Journals*? Alongside these questions are others about the meteorological record: if account is taken of the imprecision (as understood today)[39] of the instruments available to White, are the daily weather patterns possible for the time of year when they were recorded? To questions of this order, even when they include items in the garden at Wakes as unlikely as *Crambe maritima* or *Arundo donax*,[40] a strong affirmative can be given: the creatures and plants recorded do inhabit the kind of territories described, and the limits of the meteorological readings (in conjunction with allied phenomena, even particular readings for each day) are acceptable for the period for which he provides data. On this basis, we might say, there is sufficient presumptive proof to make it worthwhile to extend comment—to the matter of White's skill as an observer and as a record-maker.

Two questions are important: did White identify accurately what he saw, and, once identified, was what had been observed accurately recorded? Both are of critical importance to any form of observational science, and for both we must rely on evidence provided by White himself.

At several points earlier, it has been noted from White's letters that he claimed little skill in certain aspects of natural history. In itself, that may not be much, but it suggests that he was aware of his limitations, and that is encouraging: identifications of fish should therefore be treated with more care than those made of, say, birds. Given that White was aware of limitations of this kind, however, the procedures he adopted in order to gain accurate identifications will be instructive, and—as it turns out—reassuring. Four factors are relevant. First, he was conscious of the importance of using good authorities, namely, the volumes then available by the recognised leaders in their field;[41] second, when in doubt, he called for expert advice;[42] third, when he made records he tells us what authorities he was using;[43] fourth, he was ready to admit mistakes, both in letters and in the privacy of the *Journal*;[44] and, as well as being self-critical, he was prepared to correct others.[45] Taken together, these suggest a man with a high order of integrity, fully aware of the constraints under which he was working, but purposefully doing all he could to ensure that his findings were as accurate as he could make them. What of the next stage, the actual recording of items that had been identified?

The most detailed testimony of practice comes not from the *Journal* but from a letter. In an early communication to Barrington about singing birds, White argued that since he

carried a list in [his] pocket of the birds that were to be observed: & as [he] rode or walked about [his] business [and] marked each day the continuance or omission of each bird's song: so . . . [he was] as sure of the certainty of [his] facts as a man [could] be of any transaction whatsoever.[46]

This practice—of a pocket-book underlying a letter—can be shown with little variation to occur elsewhere, particularly in the meteorological records of the *Journal*. Two examples can be found, one concerning the actual insertion of entries, the other the sequencing of entries. For the former, attention must focus on the quantity of data given for any one day. An average number of items for the daily weather record is around five, but on some days there are many more (for example, 14 August 1770, where there are ten). No doubt some of these items may have been entered along with others at the actual time the particular observations were made, but it is difficult to envisage that this would always have been the case. The tone of the *Journal* manuscript is one of considered, even leisured, care—quite out of keeping with someone rushing to his study on each and every occasion he wished to make an entry. What is much more likely is that data collected during the day were maintained in some form of pre-*Journal* (perhaps on a slate), and transferred to the formal record later, most probably each evening although on occasion the delay was much longer. Possibly, the use of a 'slate' may have led to an increase in the number of errors, and mistakes of transcription (from instruments and in identification) might be added to those of visual reading; but the use of a 'slate' was an essential stage in the collection of data. Think how one would proceed in order to complete the weather column with prose entries. Given that the number of conditions in any one day is unpredictable, and given also a finite space (the *Journal* rectangle) in which to enter the conditions, it is inevitable that the record can be made only after the weather pattern for the whole day has been observed. Only then will it be evident whether the record can be made in just two stages (say, 'dark. fog.') or whether four are necessary (say, 'Dark. sultry. sun. fog.'). Each of these entries (which are actual records made by White,[47] and there are many more) provides visual proof of White's care. Whatever the number of conditions to be reported in the space provided, his practice is to arrange the words evenly in the column. The only way this could be done would be to store data on a 'slate', and make the written entries accordingly.

In spite of assurances concerning the appropriateness of a certain procedure, errors may still occur. We can, however, be absolutely sure that if there are any they have crept into the record against White's practice, and against his conscious caution. He knew that the very best evidences for truth were ocular demonstrations, and that whenever possible such methods must be resorted to. 'The bane of our science', he remarked, 'is the comparing one anima[l] to an other by memory in order to describe it.'[48] It is a practice he took care to avoid throughout his work.

CHAPTER TEN

The Gibraltar Correspondence

An Approach to the Study and Writing of Natural History

Although White exchanged letters with his brother at Gibraltar at the time of the death of John White senior in 1758,[1] the beginnings of a correspondence devoted primarily to the study of natural history were ten years later. In some of the early letters to Pennant, White expressed curiosity on several occasions about the winter habitat of many summer migrants, and (prompted it seems by Pennant) was successful in urging his brother

> to take up the study of Nature . . . & to habituate his mind to attend to the migrations of birds & fishes; & to the plants, fossils, & insects of [his] part of the world.[2]

That John responded actively to these suggestions is shown in the 14 letters White sent to Gibraltar. Each letter, from the first of 4 and 7 September 1769 to the last of 9 March 1772, is rich in detail about the study of natural history. Much of this material is of the greatest importance for understanding White's progress towards *Selborne* (1789)—and to that most attention must be given—but the letters also reveal White executing his role as head of the family. This family solicitude concerns two matters in particular.

In summer 1769 White received news from John that his wife, Mrs John White, was to visit England and bring with her their son, known subsequently as Gibraltar Jack, to begin his formal education, White himself having already agreed to act as guardian. Although this appears to have been the primary purpose of the visit, Mrs White was, however, also charged with making representation to relevant dignitaries concerning her husband's future. John, it seems, had decided that ten years' service at Gibraltar was sufficient to enable him to take stock of his position, and he now sought preferment at home.[3] 'It would be a very sincere satisfaction to me', wrote White, 'to see you settled again in England: but not 'til you had got an equivalent to y[r] present establishment, which has availed you so much.'[4]

On both these matters, the education of his nephew[5] and the regard for his brother, the letters show White in a role not seen in any of the documents discussed earlier. The *Kalendar*, the *Flora* and the *Journal* each declares a field of reference initiated by White himself; in the letters to John when he was at Gibraltar, the subject matter lies at the discretion of his brother. Further, the medium is different. What is dominant in any of the modes he had pursued since 1751 is the record of accomplished fact, whether that is a report of activities in the garden, the collection and identification of plants, or the events of the natural calendar. In this sense, White's various journals, although written to furnish data for future comparative analysis, are retrospective. With the letters to John this is not so. Instead of a report, there is a negotiation; and, instead of what has happened in the past lying at the forefront of his mind, it is the future. Of course, fragments of what should be termed 'report' occur in most of the letters, which usually include a brief account of Gibraltar Jack's health and education as well as mention of what progress had been made in connection with John's wish to return to England,[6] but the centre of the letters is natural history. This falls into two distinct, but related subjects: first, White's response to specimens sent to Selborne by John; second, advice about how John should proceed in his studies.

Specimens

Specimens (and various journals) arrived from Gibraltar in each of the four years 1769–72. The first cargo, which reached Hampshire on 26 August 1769, was modest and comprised no more than the skins of a few birds, one or two insects and some fossils. Later cargoes were very much more substantial and it was not unusual for White to receive twenty to thirty items on a single occasion. Many of these were birds and insects, but one collection included over thirty fish and the last collection that White mentions (which reached London in spring 1772) included a tub of plants.

At first sight, such a variety of materials from a near-alien environment would seem of little relevance to the concerns of a country parson in southern England; but as one reads White's responses it becomes powerfully obvious that the specimens and journals acted as the strongest possible incentive. 'Procrastination', he was to remark in a letter of 1792, is an 'evil power . . . which fascinates in some measure the most determined & resolute of men';[7] yet, although he claimed to have been in thrall to the 'Daemon' throughout his life, when faced with materials which at least in part were sent in response to his own prompting, his skill in natural history and the best of his temperament united to provide a concentrated demonstration of method. In the leisured pages of *Kalendar* and *Journal*, detail of how to pursue the works of Nature is scattered, needs collecting together, and is easily misrepresented. In the letters to John we can see White at his most acute; consider this:

London: June 16. 1770.

Dear Brother,

When I came to town last monday I found that y^r box of curious birds, insects, &c: had been at my Bro! Thos house about a month. In the first place y^r ruscus is the ruscus hypophyllum foliis *subtus* floriferis nudis: it is a variety . . . It does not grow wild in England.

I am much delighted with y^r specimens of ye winter-swallow, which is undoubtedly an Hirundo riparia: but tho' Lin: mentions 12 species of Hirundines, & Brisson (a french modern ornithol:) describes 17 species; yet I can make none of these descriptions answer in a satisfactory manner to y^r birds: It is with great caution that I would pronounce any animal a nondescript: & I must moreover compare it very carefully, when I get home, with our Hir: riparia: however let it prove what it will; the circumstance of it's spending the winter months *only* with you is a curious & new fact in Nat: Hist: . . .

In order that y^r insects might be properly honoured, I got Mr Drury (who is the greatest collect^r of Insects in this kingdom) to open them, & to set them in boxes. He was much pleased with several of them . . . He says y^r situation is a very good one for insects, & that many curiosities may be found with you. He desires you would attend to ye following directions . . . I must endeavour to get an Icthyologist at the opening of y^r fishes, which I am not so well acquainted with as I am with birds. You seem to make a very quick progress in Insects: attend to the generic distinctions & you will soon read them all . . . When I have considered y^r curiosities better, you shall hear farther . . .[8]

In these few lines there is distilled the epitome of White's approach. Most startling perhaps for a reader who may know only *Selborne* is the directness. Rather like the entries in a journal, the preliminaries are confined to a minimum—the brief citation of place, date, and circumstance—and attention is given immediately to identification, and to the enthusiasm of the search. The specimen of butcher's-broom[9] gave little difficulty, but the 'winter-swallow' taxed him to the utmost. As White says, the bird resisted his resources in London and he therefore gave it a provisional name applicable to the time of its known appearance: hence, *Hirundo hyberna* ('winter-swallow').

That White was unable to identify the bird is of greater interest than it might seem, for it provides an opportunity to watch him at work. The field of inquiry (ornithology) is one in which he excelled, but the procedure he adopted applies also to other aspects of natural history and can therefore be taken as exemplary, especially as it demonstrates one of White's great strengths, as well as what the world might count a weakness.

Faced with a bird that was unidentified, except that it was a kind of swallow,[10] White did two things: taking the several specimens back to Selborne, he first sought to establish its relation to the swallow that nests in sandy banks in England, *Riparia riparia*, sand martin. He did this on 12 July by comparing the specimens from Gibraltar with birds he caught himself that evening locally.[11] Second, he sought advice — from Thomas Pennant, and from William Sheffield (the future provost of Worcester College, Oxford). To each he gave a specimen and hoped that where he had failed to establish an accurate identification they would be more successful. As things turned out, however, it was actually White himself, albeit with some doubts, who established what the specimens really were: examples of *H. rupestris*, crag martin. To seek confirmation of this identification, he sent both Pennant and Sheffield a copy of a description of the crag martin that he had discovered in an account of the birds of part of Austria published by John Scopoli;[12] some features of the description by Scopoli were at variance with the specimens White had to hand, and it seemed circumspect to obtain confirmation of his identification before proceeding further.

How Pennant met White's doubts is not on record (we know only that he accepted the identification), but Sheffield's response is extant and it is from that that we can perceive White's weakness: a lack of grasp of formal, systematic ornithology. In response to the description from Scopoli that White sent, Sheffield wrote:

> . . . from the Description there seems to be no doubt but [Scopoli's] rupestris is your Brother's [bird]: I lay little stress on those passages You have marked as exceptionable, having observed in Linnaeus & other Systematists that the specific Difference is almost universally constituted by one, or two particular Marks, the rest of the Description running in general Terms — some thing like the Expression *more* or *less*, in our own Language, when referred to Estimates . . .[13]

What, of course, this provides is a radically different understanding of natural history. White was a practical, field naturalist: for him, the eye was superior to the book, and, having compared the specimen in the hand with the verbal description, he had found the latter weak in detail. For Sheffield, accustomed as he was to the theoretical language of academic debate and proof, it was understood that the weighting to be given to the parts of a verbal description could vary, and that, so long as there was identity between description and specimen in respect of an unambiguous feature ('the specific Difference' — in this instance, the oval white spots on some of the tail feathers of the crag martin), then the remainder could be read at a very general level.

To White's credit, he accepted Sheffield's stricture, and much later incorporated the judgement in *Selborne*:

> Foreign systematics are, I observe, much too vague in their specific differences; which are almost universally constituted by one or two particular marks, the rest of the description running in general terms.[14]

The nature of the judgement offered, however, is noticeably different. Instead of remarking on how systematic descriptions are to be understood (which is what Sheffield had done), White condemned them outright. In fact, the mode adopted by Linnaeus was much in advance of anything attempted earlier, and White's failure to appreciate the economy and elegance of the new style was not only a decided weakness—for it restricted his ability to make best use of texts written in the mode—but also hindered his appreciation of the direction in which the study of natural history had begun to move.

If we pause to assess what White had achieved so far, these limitations are unexpected since they appear in his work at a time (1770–1) when he was at the height of his powers. His interest in nature for over thirty years had been extended into the serious studies that had culminated in the *Flora*; the notion of a natural calendar had been embraced with enthusiasm and conviction, and records for several years had been maintained in the *Journal*; his knowledge of natural phenomena was such that it was valued by the leading writer of the period, Thomas Pennant, with whom he had pursued a voluminous correspondence for three years; he had been introduced to Joseph Banks, who had appreciated the receipt of specimens of what must, to White, have seemed relatively common productions; the Hon. Daines Barrington, with whom he was also corresponding, thought sufficiently highly of his skills not only to advance them by introducing him to ornithological collections in London,[15] but also to encourage him in the preparation of a fauna of Selborne; and his own confidence was such that he was prepared not only to invite naturalists to Selborne to show them the phenomena of the parish, but also to tutor his brother at Gibraltar in the pursuit of natural knowledge and to encourage him to correspond with the most distinguished naturalist of the age, Linnaeus.[16] Moreover, this level of accomplishment was sustained over several years and up to the presentation of papers read at meetings of the Royal Society in the springs of 1774 and 1775. We must conclude that, although he never mastered the lesson afforded by Sheffield's explanation of how to read a description prepared in the new fashion, there is little need to feel that White is thereby diminished. Today, the volumes of descriptions White took such exception to are used only by a few professionals; whilst his own *Selborne* is justly celebrated, and still read. In addition, there were other compensations over and above those qualities of directness and enthusiasm with which he began his June 1770 letter to Gibraltar. The most conspicuous is his candour, although in the long run, and if *Selborne* is taken into account, his triumph over procrastination is as important, if not more so.

Mention of White's integrity has been made earlier in connection with the deletion and correction of certain entries in the journals, but it has several facets. In addition to acknowledging error in oneself, there is the willingness to acknowledge doubt and uncertainty to others. This, of course, is what White does in seeking advice not just to confirm an identification (*H. rupestris*), but to bring credit on the collector of the specimens ('In order that y^r insects might be properly honoured . . .') and to initiate identifications ('I must endeavour to get an Icthyologist at the opening of y^r fishes . . .'). But the importance of integrity extends further than this, to the very act of writing with a view to publication. As White had responded to Barrington when the latter first suggested that White himself should write: 'investigation (where a man endeavours to be sure of his facts) can make but slow progress'.[17] This kind of candour, a regard for the accuracy of what one writes, was an attribute he used to assess the merits of an author, and one that had predisposed him as early as January 1769 in favour of Pennant (who readily admitted error) as against Linnaeus (who, somewhat self-consciously—so White thought—apologised for making changes in his writings).[18] It was something his brother would have to learn to seek advice about, for 'slips, & unavoidable errors'[19] would make their way in a man's work however careful he was.

But this is to trespass on advice he was to give John, and before we approach that there is another aspect of White's integrity, namely his persistence, discussion of which is also prompted by the June 1770 letter. 'When I have considered y^r curiosities better', he had written, 'you shall hear farther.' Yet, more protracted attention does not always yield an answer, any more than it did in connection with the migration of swallows and martins.

The most instructive, and sustained, example in the letters concerns *Nemoptera bipennis*, a distinctive variety of lacewing found only in the Iberian peninsula. White's introduction to the species was as an item in the box of specimens brought to England by his sister-in-law in 1769. Immediately, he recognised it as unusual and, although it was identified the following year, more specimens were requested since it was 'a curious rare insect'. Apart from the rarity of the creature, White was intrigued by its remiform (oar-shaped) wings, and as well as urging John to discover its life-cycle he enquired whether the wings were of assistance 'in copulation'. Unfortunately, in spite of persistent prompting throughout 1770–2 John was able to offer few answers. White, however, was undaunted and when John was resident at Blackburn in Lancashire he was still urged to make enquiries about the insect among friends in Gibraltar with whom he maintained communication.[20]

No doubt part of White's tenacity derived from the interest shown in the insect by other naturalists—Lee, Drury, Sheffield, and Linnaeus, all applauded the discovery—but that alone seems insufficient to account for it. That John had been unable to find answers to his brother's enquiries was no failing: modern naturalists have also not established full details of its life-cycle, and, although it is known that adults are pollen feeders in an arboreal habitat of pine and eucalyptus, the function of its remiform wings (each of which shows a spiral twist asymmetric with the twist on the other wing) is unknown.[21] In such circumstances, one must look for explanation elsewhere. It is this: in urging his brother to seek more data, White reveals the cast of his own mind, in two senses. First, he had no time for the shallow, incomplete response: if something was worth pursuing, then persistence was likely to be its own reward; and second, although the example of the crag martin may have suggested that what White sought was an identification, what he was most interested in was 'ye life & manners of animals [which] are the best part of Nat: history',[22] and his persistence over *N. bipennis* tells us just that.

These examples, of White's directness, enthusiasm, candour, and persistence, are not confined to the few specimens already discussed. Throughout the letters to Gibraltar, the primary focus is on the curiosities that he received and shared with others.[23] Such was his pride in what John was accomplishing that he did what he could (apart, that is, from assenting to a suggestion of John's that live creatures should be sent to England)[24] to publicise everything that arrived. Living as he did in a quiet corner of Hampshire, he knew few could benefit from the specimens if they were kept at Selborne. Initially, then, as well as sharing specimens with friends, he arranged for the cargoes to lie 'in London for the inspection of the curious' before being sent down for him to study; and later he determined: 'Where I have duplicates I shall present them to some Gent: of fortune that are making collections' for public display. On how many occasions such presentations were carried out is uncertain, but for one notable instance documentary evidence survives. Writing to John in March 1772, White determined:

> Your Flamingos are so rare, & so much admired, that I shall give one
> to the musaeum of the royal society, & the other to a Gent: of fortune
> who is a great collector; chusing that they should belong to places
> where they will be set in natural attitudes & much admired; rather than
> that they should be laid up in a box in my study where they will seldom
> be seen.[25]

And in the records of the Royal Society, under the heading 'Presents
made to the Royal Society in the year 1772 . . .', is the note:

> Mar. 26. Mr. White. A fine Specimen of the Phoenicopterus, or
> Flamingo.[26]

Advice on the Pursuit of Natural History

White's tuition of his brother in the pursuit of natural phenomena had
begun in 1768, and the 1769 cargo of insects and birds was the first
tangible result. However, the earliest known letter addressed to John
at this time, that of September 1769, makes mention of a meteorological journal. No doubt this is the kind of document referred to by
John in a letter to Linnaeus of June 1771, where he claimed 'for some
years [to have] kept a journal of the winds and weather, and
variations of the thermometer and barometer'.[27] Almost certainly
such a journal was kept at White's suggestion; and taken with his
other requests (all before 1770)—that John 'keep a Flora . . . an
Hortus siccus' and that he should 'Describe what strange fishes you
see . . . Collect shells on y^r shore; & any nat: fossils'—and John's
known beginnings in the collection of insects and birds, it adds up to
a programme that is remarkably unselective. That John was expected
to work up all the varied facets of natural history simultaneously is
shown also in the books which White arranged to be sent to Gibraltar
in 1769.[28] As well as a natural history of Minorca (sent to show 'what
has been observed in a neighbouring Island: that You may not
advance any thing as new that has often been observed before') and
an account of 'the manner in which the disciples of Lin: investigate the
nat: curiosities of a Country', texts were provided on each of the main
divisions of the subject: birds, fish, insects, plants, fossils, mineralogy.
In some instances two different texts were provided and, where he
could, White recommended books that were apposite because of the
latitude in which John was working. How far the source of these
various recommendations was White's own is not obvious since
some specialised volumes came from Pennant, who, once he had
heard about John's pursuits, was very keen 'to promote them'; this

was particularly so in connection with the fish that John might discover, and he not only advised White about which books his brother should read but also made gifts himself. Whether his motive was as disinterested as White's is doubtful,[29] but the encouragement it afforded both White and his brother should not be ignored.

To ask the student of today to pursue a study of natural history at the forefront of knowledge as inclusively as this range of text-books suggests would be unusual. Yet this is just what White expected of John, although his own quite recent experience of serious study (work on botany in 1766) might have guided him towards a more phased approach. Nevertheless, that an observer of the natural scene of White's experience expected a novice to begin his studies on such a broad front all at once (and as a form of distance-learning, with no support from Open University broadcasts) tells us much both about White and about his brother, and also something about eighteenth-century attitudes to the subject.

An obvious place to begin is with the high expectation White must have had of his brother's capabilities; and to judge from John's known writings and from the opinions of Mulso and others any doubts we may have, based on the non-publication of the results of his investigations, are probably falsely derived.[30] White himself was enthusiastic in support, and consistently so, and that in itself must count for a great deal. But two other points emerge—one general, one personal: both are considerably more important.

The general point concerns the largely undifferentiated nature of natural history at the period. This, of course, controlled the kind of study that was possible. Predominantly, the climate of the age meant that the focus of work was on identification of species. This is well illustrated both by the work of Linnaeus and by the displays and museums created to house the huge collections brought back to England by travellers from overseas. A good example of this latter is the display exhibited by Banks following his circumnavigation: Sheffield's account, in a letter to White, shows not only his amazement at the sheer quantity and variety of the collection, but also the range of what natural history then embraced, namely

an almost numberless collection of animals; quadrupeds, birds, fish, amphibia, reptiles, insects and vermes [snakes] . . . 987 plants . . . and 1300 or 1400 more drawn with each of them a flower, a leaf, and a portion of the stalk. . . . And what is more extraordinary still, all the new genera and species [on Sheffield's evidence 1300 new species] . . . are accurately described, the descriptions fairly transcribed and fit to be put to the press.[31]

With such a wealth of material pouring into Europe it was inevitable that naming and description were priorities: until a sufficiency of resource had been accepted universally (and the maintenance to this day of specimens collected by Linnaeus and Banks as standard reference specimens bears witness to the importance of the task) the relevance of other modes of activity was forestalled.

Further, in addition to the need to classify all the newly-discovered creatures and plants, there was an absence of disciplines centred on particular forms of nature. There were some specialised books, but this was primarily a matter of convenience and the necessary progress of natural history (some would say its inevitable fragmentation) into the establishing of separate subjects, for example, ornithology, herpetology, and others, did not occur until the next century.[32]

But if White encouraged John to pursue the whole of natural history because that was in accord with the times, there were more personal reasons at work also. Earlier, mention was made of the enthusiasm, directness and persistence shown in the letters White sent to Gibraltar. Such qualities might be taken as signs merely of a brisk efficiency: in fact, they go much further. The key to this is found in a letter of October 1771. Of all the kinds of classification John's studies led him to engage in, he came most slowly to plants. The difficulties he experienced are referred to several times by White, who repeatedly offered encouragement. At the time of the 1771 October letter, White had had a guest from Oxford (Richard Skinner of Corpus Christi College) staying with him who had talked with White's nephew about Gibraltar:

> Y^r son Jack [reported White] got great applause from Mr Skinner by the satisfaction he gave him respecting the manner how the dwarf palm carries it's fructification: you will please to send some of the fruit, an entire spatha as it grows.[33]

Praise the son, White must have thought, and thereby charm the father! That is not all: although consumed by renewed enthusiasm, one's knowledge may be such that advance is inhibited, yielding, of course, frustration rather than success. White saw this also, and added to the passage cited more guidance:

> If I could step to you now & then for a day, I could set you forward in y^r botany, tho' I am no great adept. The terms are formidable & forbidding at first, but easily learned. Lee's [*An Introduction to Botany*] is an excellent introduction.[34]

Although John's difficulties with botanical study must not be attributed to White's not visiting him, the notion that White might take a 'step' to Gibraltar 'now & then for a day' tells us much. Taken in conjunction with the spirit of the letters as a whole, White clearly obtained enormous satisfaction from his brother's progress and achievement. Indeed, so great was the satisfaction that, when he came to explain about writing up the investigations, his guidance took the form of a full prospectus.[35] After commenting on an appropriate dedication and preface (which should remark on how fitting it was for a clergyman to study natural history 'when considered in a physico-theological light'), he then guided him concerning the need for a 'clean, easy, & correct' style and proposed that the main work should be introduced by a civil history of Gibraltar, details of which White himself proposed to obtain from Barrington;[36] and that the substance of the work should comprise three elements: description of species, a journal of one entire year, and 'dissertations'.

The idea of description of species is plain enough, but both the notion of a journal and that of incorporating dissertations require comment. These last were understood as essays on some aspect of John's experiences at Gibraltar, and their inclusion was advised because otherwise 'you will lose half y^r readers, who know nothing of system, & zöology'. White encouraged two topics in particular. One was of long-standing interest to shipping: how to negotiate the complexities of the tidal flows at Gibraltar between the Mediterranean and the Atlantic.[37] The other was an essay in anthropology based on a visit John had made across the Straits:

> Make a narrative of y^r voyage to Morocco independent of Nat: hist: describe the person of the Emperour, his manners, guards, troops, discipline . . . Describe the people of Gib: the ancient Spaniards, &c: &c: the population, & means of depopulation; what proportion births bear to deaths; the different effects of different winds: the E:, I find, generally brings fog. Say some what of the Jews; if those of Barbary differ from those of Europe . . .

Matters such as these were not, however, to form a separate part of John's work: 'You must mix with y^r nat: descriptions now & then a dissertation', White had written.

Descriptions and 'dissertations' of the kind suggested were matters about which White could do no more than advise; they had to be based on observations John alone was able to make. A similar restriction would seem to apply to a journal of meteorological data, but it is with this proposal that White reveals the very heart of his interest. Reference is made to the proposal throughout the letters, but the leading features occur in two letters of 1770. 'Publish [he wrote in August, see Plate 7b] a journal of one entire year with its coincidences'; then, prompted by the example of what Stillingfleet had offered in the *Tracts*, he enlarged:

> Mr Stillingfleet, you see, has opposed the calendars of Upsal & Stratton. Now Sweden is about as far N: of England as Gibraltar is S: therefore (if you approve of it) I can furnish you with a diary from Selborne for ye year 1769: by which means these three climates will be exactly compared, so as to afford some information to a curious naturalist. My Journal will be very full & particular . . .

A radical weakness in this idea was soon pointed out to him (perhaps by John): that to compare data of 1755 (the Stillingfleet data) with those collected in different countries for 1769 was thoroughly unsound. At any rate, within a month or so the proposal was modified to the more feasible:

> Pitch on some one year for y^r Journal suppose 1769: & then throw all the current observations you have made, & regular incidents into that year: by which means there will be some times 4 or 5 pages of observations to one of Journal. In that manner I shall manage my Journal for the same year: & thus we may compare the two climates.

Clearly this is a more realistic proposal, but if we are to understand exactly what White intended some explication is necessary, particularly in connection with the distinction that must be drawn between on the one hand what is appropriate in a 'Journal' and on the other hand what is meant by 'current observations . . . & regular incidents'. White is writing, we need to recall, in 1770, so anything that is 'current' must relate to that year; and yet the 'Journal' is intended to record events in the previous year, 1769. The nature of the difficulty appears in its sharpest form in an earlier proposal of August where White had written:

> Publish a journal of one entire year with its coincidences: [and continued] & where incidents *naturally occurring* escaped you in that year; you may fairly insert them from a preceeding or following Journal.

Momentarily, this seems remarkably inept as a scientific proposal: failure to observe a certain phenomenon in one year cannot be hidden by inserting it in the record for another just because one knows it is *naturally occurring*. We expect better of White than that, surely? Indeed yes; and we are given it. The problem lies, of course, in the nature of natural history events: although they are observable, they are not predictable in the way that such things are in the world of the exact sciences; moreover, some expected events may actually not occur in a certain year.

Discussion of some relevant approaches to this problem was pursued earlier in connection with reading White's own journal entries. At that stage we were concerned with, for example, the kind of accuracy his instruments permitted, levels of competence of transcription, and sheer knowledge of species: to those must now be added a quite different dimension. Previously it was suggested that White's belief in the value of participating in the programme of observation permitted by Barrington's *Journal* format began to wane during the early 1770s, and that the entry of data in columns gave way to a form of record-keeping that was more reflective. If we bring to that suggestion this proposal of White's to his brother concerning a journal for 1769, we can see that as early as 1770 he had realised the limitations of the journal for recording everything that a naturalist *should* observe— if, that is, one were available for 24 hours of each day throughout the year, if everything occurred that prediction had led one to expect to occur, if seasonal change embraced normal averages as well as extremes all within the same year, and so on. Obviously, to ask for this is to ask that natural history be not what it is, but something else; and so a different solution must be called for. This is precisely White's proposal. John had a journal of meteorological data and other natural phenomena that he had observed in the year 1769; many occurrences that he had commented on previously or observed subsequently were missing, but a place must be found for them in a publication purporting to provide information about the natural calendar at Gibraltar. The solution, therefore, was for the actual journal for 1769 to be accompanied by a separate section which reported 'observations' that *customarily* occurred at the season declared. Sanction for the kind of entry White had in mind had appeared in Stillingfleet's *Tracts*, a representative entry reading: 'BIRDS OF PASSAGE, after having celebrated their nuptials in the vernal months, and feasted on the summer fruits, now prepare for departing.'[38] An entry of this kind would be classified as one of the 'regular incidents' to which White refers in his proposal and on a published page would be presented in a typographically different form from actual dated observations for the given year. As for the quantity of material White expected John (and himself) to produce, the dated Journal for 1769 was to be accompanied by four to five times as many observations and incidents drawn from other years.

This notion, the study of events '*naturally occurring*', had a profound effect, not just on the continuing appeal of what White was later to write, but also on how he approached his own eventual *Selborne*. Indeed, the advice he gave John is interwoven with his own movement towards becoming an author. For some months he seems to have envisaged a collaborative endeavour of the kind suggested in his proposals to John, but at some point he took the decision to write independently. An important moment in that process occurred in response to the 1770 guidance he had offered his brother; but other factors also contributed, and it is proper to give separate attention to this new development in White's thinking—the movement towards his own publication.

CHAPTER ELEVEN

Towards *Selborne* (1789), and After

The origin of *Selborne* is usually taken to be a suggestion made early in 1770 by Daines Barrington that White should prepare 'an account of the animals in [his] neighbourhood'.[1] In the light of what White had by this time achieved privately and for the most part informally, the idea was probably disconcerting, but it was certainly a positive factor in his eventual recognition of his own capabilities. If we compare the suggestion with the final outcome (*Selborne*), we can see a considerable difference between the two. It will, therefore, be the main purpose of this chapter to trace the development of White's response to this first encouragement that he should become an author, from the hesitancies of 1770 to the accomplishment (and reception) of what he finally wrote; and in a final chapter to offer an assessment of *Selborne* itself.

Deciding What to Do: 1770–1774

White's initial response to Barrington's suggestion was favourable. 'When we meet', he wrote on 12 April 1770, 'I shall be glad to have some conversation with you concerning [your] proposal'. Yet, at the same time, he expressed some caution about his ability to match the encouragement he had received:

> Your partiality towards my small abilities persuades you, I fear, that I am able to do more than is in my power. For it is no small undertaking for a man, unsupported & alone, to enter upon a natural history from his own remarks. Tho' there is endless room for observation in the field of nature, which is boundless; yet investigation (where a man endeavours to be sure of his facts) can make but slow progress: & all that one could collect in many years would go into a very narrow compass.

It is possible to read these reservations as indicative of a needless modesty, but a much more fair judgement is to acknowledge them as honest self-appraisal. As White had learned at Selborne the previous year from the brief companionship of Skinner and Sheffield, visitors could be more adept than he at identification; and further, committed as he was to the importance of an absolute veracity, the serious studies he had undertaken over the past four and more years had shown that 'the field of nature [is] boundless' and that it took a great deal of time to collect any worthwhile data. More importantly (and this is a mark of White's originality), the kind of account he would wish to write would not be a volume of mere names, descriptions and synonyms (important though a systematic approach might be), but one, as we have seen in his advice to John, devoted to the 'life & manners of animals [which] are the best part of Nat: history'.[2]

White wished to pursue his studies further; he had opportunity to do so, and he had taken a decision about what kinds of observation were most worthwhile. There yet remained the matter of geography: within what kind of region should data be sought? In this sense White must have rejoiced at Barrington's proposal. Selborne had been the canvas on which he had made his observations from the time he had first looked about him; it had been the beginning, and (whatever fascination was afforded by acquaintance with what occurred elsewhere) it was to be the end, of his experience. Now he was being invited to write about what he knew, not in private correspondence but in a form designed for the public. His delight must have been spacious.

The narrow focus brought about by the apparent restriction of territory designated by the term 'neighbourhood' was, of course, an inevitable consequence of White's holding to the importance of observational truth. If one wishes to be sure of one's facts, familiarity with what to expect and where to go in order to find it become necessary conditions of verification and progress. Certain of this conjunction, White rapidly turned it to his advantage and claimed (what history has confirmed as a further mark of his percipience) the importance of local studies. Writers of his time had for the most part encyclopaedic hopes, endeavouring to document territories, if not as vast as the known world (Linnaeus's intention), at least as large as a whole country. But one writer whom he had discovered (Scopoli) was publishing on a smaller scale;[3] and White skilfully adopted him to his cause. As he wrote to Pennant in 1770:

> Monographers have a fair pretence to challenge some regard & approbation from ye lovers of Nat: history: for as no man alone can investigate all the works of nature; these partial writers may, each in their different department, be more accurate in their discoveries, & freer from errors than those that undertake in a more general way; & so by degrees may pave the road to a correct universal Nat: history.[4]

With views as firm as these (and they were addressed to Barrington also, in October the same year),[5] we might be forgiven for thinking that White's 'conversation' with Barrington would yield a more rapid result than it did. White himself was in London in June 1770, and the personal interest must have been insistent: he met Barrington on several occasions and rejoiced at being shown many of the finest collections of natural history specimens then on display.[6] He must have felt indebted.

White was not, however, of a temperament to be precipitated into action, and over six months were to pass before Barrington received a considered reply. The reason for this delay was neither a lack of interest, nor a reluctance to declare himself, cautious though he might be at the prospect; rather, it should be seen as prognostic of the (almost) two decades that would be needed to bring any plans for writing to a successful conclusion. Of course, these years were not to be devoted exclusively to the implementation of Barrington's pro-posal—that was to evolve through several stages before it took a final form—and there were always many other necessary things to attend to. Chief among these was his absorption throughout 1770 in corres-pondence with Gibraltar. In May he had written a long letter of advice to John, and when he went to London in June he had found awaiting him the important box of curiosities that included the 'winter-swallow'. Arrangements had to be made for inspecting the specimens of insects and fish, and decisions sought about the best way to encourage John and to promote his work. This was not all. As soon as he returned home at the end of the month he was needed at Fyfield, the home of his brother Henry, which he was to visit again in September; Mulso and family were to stay with him in August, as also was Sheffield; and there was a journey to make to Ringmer to see his great aunt, which was to take nearly three weeks out of October; clerical duty to be discharged at Faringdon; and the garden to oversee at Wakes. In the midst of all this, almost before he had assimilated the specimens that had arrived in June, a further cargo reached him; his embarrassment was complete:

> I am now to thank you [he wrote to John on 6 November] for a most curious collection of birds & insects, which I received about the last week in August: & the reason I did not take any notice of them before by letter was, that thro' the following months of Sep: & Oct: I was much taken-up with company & Journies, & could not pay them the regard they deserved.[7]

The 'regard they deserved' derived not only from the intrinsic interest of the specimens themselves but also from recognition of responsibility. John had launched himself into natural history largely at White's request, and it is a fair reflection of personality that (at this time particularly, during the early stages of inquiry) he should place John's needs before his own.

Given this acknowledgement of priority, it is easy to foresee that, when (on 12 February 1771) White did compose his considered response to Barrington's proposal, it would contain as leading ingredient an abbreviated form of what he was guiding his brother towards:

> In obedience to yr repeated injunctions [he wrote] I have begun to throw my thoughts into a little order, that I may reduce them into the form of an annus historico-naturalis comprizing the nat. history of my native place.[8]

Diffident though White was about his ability to bring such a proposal to fruition, the influence of the calendars published in Stillingfleet's *Tracts* and of his own completion for over three years of the *Naturalist's Journal* (behind which stood the 17-year cycle of the *Garden-Kalendar*) is obvious. More importantly, how best to present to the public a *local* natural history had been at the front of his mind in the advice he had been sending to John at Gibraltar, and, as has been seen in the preceding chapter, White was enthusiastic about a journal composed of data observations to which were added necessary 'co-incidents'.

An *annus*, however, need not present information in calendarian form: data might be organised, as Scopoli had done for example, into broad categories in which the dating of phenomena is almost ignored; although a dated format must have been White's intention. Confirmation comes from his pen on two occasions. The first is in a letter to John written only a few weeks before he made his proposal to Barrington:

> As matter flows in upon me I begin to think of composing a nat: Hist: of Selborne in the form of a journal for 1769 . . .[9]

There can be little doubt that the form of this journal would be that described in the advice given to John which we have discussed earlier: that is, a dated journal with 'co-incidents'. The second occasion that can be given as evidence of the form White had in mind is much later, at the time the writing of *Selborne* and the 'Antiquities' was formally concluded. The last letter in *Selborne*, written above the date 25 June 1787, begins its final paragraph with these words:

> When I first took the present work in hand I proposed . . . an *Annus Historico-naturalis*, or The Natural History of the Twelve Months of the Year; which would have comprised many incidents and occurrences that have not fallen in my way to be mentioned in my series of letters . . .[10]

No firmer indication of White's enthusiasm for the cycle of the year could be found; and in part our task now must be to account for its absence in his eventual volume. In *Selborne* White explains that an *annus* was not included because another author had recently published something similar. That, of course, is an explanation which belongs to 1787; it scarcely accounts for the existence of *Selborne* itself—which by then afforded him the privilege of excluding the proposed *annus*.

How then did *Selborne*, as we know it, come about? The answer must be found in White's activities between the proposals he made to his brother and to Barrington (in January and February 1771 respectively) and what is a recognisable prospectus for *Selborne* which he wrote in 1774. Over and above a general sense of confidence derived from what he had accomplished since, say, 1766 (the year of the *Flora*), there are several other developments that can be adduced.

One concerns the naturalist Richard Skinner. After visiting Selborne with Sheffield in 1769, he returned in September 1771 (the occasion of his praising the description given by Gibraltar Jack of the flowering of the dwarf palm). In response to some mention by White of his plans, Skinner encouraged him to write about the subject he knew best—birds:

> M[r] Skinner [wrote White in autumn 1771 to John] exhorts me much to publish an history of the passeres of Selborne. We abound in summer birds of passage: I can recount 20 species.[11]

The merit of the exhortation is more significant than it may seem. History records little about Skinner himself, but from correspondence we know that he was on good terms with some of the foremost natural historians of the day, and familiar with many current activities. Evidence for this is given in letters White received both from Skinner and from a man who was to achieve scholarly acclaim in the field of botany, John Lightfoot. Together, these colleagues (for both were Oxford men, and in this regard we can add the contribution made by Sheffield) kept White in touch with some of the most important events in the field of natural history almost as they happened. This is particularly true for Lightfoot. In 1772 he accompanied Pennant on a well-documented tour to Scotland[12] (resulting in his *Flora Scotica*, 1777) and followed this in 1773 with a tour, this time as companion to Joseph Banks,[13] through the whole of Wales. After each journey White received an account of Lightfoot's main impressions, which embraced not merely comments of a botanical nature, but observations of a general kind also.[14] In such a fashion, apart from his continuing and direct correspondence with Pennant (and Barrington), White maintained an understanding, if not of the theoretical advances in his subject, at least of the incidental and topographical, and this was to have some importance almost immediately. But before turning to that there are important developments in two other areas to be considered: one is public—at least there is a public outcome—the other familial.

The public outcome centres principally on Barrington. Pennant, however, also made a contribution, and in a way that revealed the strength of White's temerity and authority over matters that derived from observation. In autumn 1773 Pennant wrote to several correspondents, seeking further information for a new edition of his *British Zoology*. White's response was a long list of corrections and additions; but he was very conscious of the service he was rendering and concluded his remarks with this apology:

> On a retrospect I observe that my long letter carries with it a quaint, & magisterial air, & is very sententious: but when I recollect that you requested stricture, & anecdote, I hope You will pardon the didactic manner for the sake of the information that it may happen to convey.[15]

If the tone of this is not sufficiently clear, what White privately intended is made explicit to his brother:

> I have received [he wrote to John] a most violent complimenting letter from Mr. Pennant lately. He is going to publish a second edition of 'Brit. Zool.,' and is to do wonders with the information extracted from my letters. I shall take the opportunity of laying before him the more glaring faults in the first edition.[16]

White is clearly gratified, but amused. Yet, although Pennant was quickly satisfied, the impetus White derived from receiving and meeting the request should not be minimised: as always, his weak intrinsic motivation was stimulated into controlled activity because of the external pressure.

To meet Barrington's requests, however, considerably more resolution was necessary; but the outcome was commensurate. From the beginning of their acquaintance (and even before), Barrington had sought to enlist White in the advancement of learning and knowledge: recall, for example, not only the original gift of the *Journal*, but also the promise he obtained from White in 1769 to prepare a *methodus* of singing birds and birds of passage, as well as the 1770 proposal that underlies the present discussion—an account of the animals of his own neighbourhood. How frequently, after White's response of February 1771, Barrington enquired about progress is uncertain, but correspondence flowed rapidly from one to the other and it seems likely that he may have judged, if a beginning was to be made, that a further initiative might be advantageous. At any rate, sometime in

1773 (possibly during White's spring visit to London) Barrington persuaded him to write a set of essays on his favourite creatures, the members of the swallow family. After a summer of close observation and verification of existing knowledge, the first monograph, on the house martin, was completed on 20 November of the same year and within a few weeks White was able to write to John:

> My monography on the house-martin is finished, and in the hands of Mr. Barrington, who is so much prejudiced in its favour, that he proposes soon to have it read before the R. Society. Another on the house-swallow is nearly completed; from which I propose to proceed to the rest of the British *Hirundines*.[17]

Barrington was as good as his word, and White's first paper was read at a meeting of the Royal Society on 10 February 1774. The reception was favourable, and within a month White was in London. It was a timely visit.[18] On behalf of the Royal Society Barrington had recently arranged for the Society to exchange natural history specimens with the King of Spain (the Society was to furnish materials from North America, the monarch specimens from South America), and White, accompanied by Gibraltar Jack, who had come with his uncle to London, was conducted through 'the curiosities of the R.S.' at the very time materials were being packed for shipment. For White and his nephew to witness these preparations must have been a rare privilege, touching as it did matters in which they had more mutual interest than would most uncles and nephews of the time.

For us today there was a far more important outcome, one based on fraternal companionship. On returning to Selborne, White sat down four days later (29 March) and wrote to John:

> As to my letters, they lie in my cupboard very snug: if you will correct them, and assist in the arrangement of my journal, I will publish. I have finished the monography of all the *Hirundines* except the swift.[19]

The significance of this can scarcely be overestimated. Somehow, to the 1771 notion of an *annus* in the form of a journal there had been added the idea of letters: letters, apparently, not fabricated for the purpose, but transcribed copies (presumably) of real letters. The implications of this for the continuing survival of the finished work are fascinating (and will be pursued later); of equal interest is the purposefulness of the comment and the calls for correction and assistance. The purposefulness is easily explained and should be seen primarily as an outcome of Barrington's patient and perceptive encouragement which had led to the writing of the swallow mono-graphs — and consequent success at the Royal Society; but the call for correction and assistance is altogether different and merits separate consideration.

The most significant aspect is the addressee, brother John, who had returned from Gibraltar in 1772. His first task had been to settle the details of his preferment (to Blackburn, in Lancashire), but once that was done he had travelled with his wife to Selborne. Arriving on 8 December, he stayed there until 26 April 1773, and much time was spent in discussion of natural history: 'I preserve all y[r] letters & Journals, that you may have recourse to them before you publish',[20] White had remarked in his last letter to Gibraltar. Further, the arrangements for the winter were a great bonus to White himself (and in ways other than the strictly scientific), so much so that he hoped his brother and sister-in-law would be able to repeat the visit the following year:

> I wish you and my sister [he wrote in September 1773], while your house drys, would come and spend one more winter with me. You might put in an elderly, grave person for that period to make fires, and take care of your goods, and defer the hiring of servants 'til spring. I will endeavour to do everything to make the winter as easy to you as possible; you shall have a bed put up in the drawing-room, and a grate where you shall have a constant fire, by which you may instruct your son, and fabricate your *Fauna*. As my sister, I know by agreeable experience, is of an active disposition, she shall, if she please, manage my house, and see to provisions; and we shall, if it please God to bless us with health, pass the dead season of the year in no uncomfortable way; and at the season of spring I will let you depart in peace, and will follow you in the summer into Lancashire. All this proposal is the result of a sincere intention; and therefore I hope you will think of it in earnest, and not let the consideration of a long coach journey . . . prevent the comfort and satisfaction I propose from such an undertaking.[21]

Inviting as this may seem, more particularly, perhaps, because Gibraltar Jack was still resident at Selborne, John demurred and instead of coming south invited his brother north to Blackburn. White found the prospect appealing and made preliminary arrangements, including a determination to make himself free to travel by resigning his curacy at Faringdon. However, for one reason and another (and in spite of repeated invitations, for White always discovered there was a more compelling priority) the visit never took place. His regret was sincere:

> For my own feelings, I often wish myself with you [he wrote to John in January 1774], and make many comparisons between this and the last winter, not much to the advantage of the present. Last winter I look back upon as one of the most pleasant of my life, when I had my

friends about me in a family way, and enjoyed the conversation of relations from whom I had been parted so long.

Conversation could continue, however, by letter, and in a style that is the most natural and episodic to be found in all his correspondence. White kept his brother fully informed about events at Selborne, family activities, progress of Gibraltar Jack, his own circumstances and all the other occurrences (inconsequential except to a biographer) that two brothers, different in temperament but close in occupation and interests, would wish to share. Reading the letters, we glimpse a relaxed mutuality of confidence and respect which suggests that White had, at last, found that long-sought-for companion capable of quickening his industry and sharpening his attention.[22]

That this is so becomes more clear if we reflect on White's asking John to become his corrector and adviser. The brother at Gibraltar whom he had inducted into the procedures of the natural historian had attained skills that matched those of his master: John could now be treated as an equal. Moreover, the authority and trust White now vested in his brother ensured that it was to him that White sent what can be termed a full and recognisable prospectus for the eventual *Selborne*. Evidently, the brief suggestion in his letter of 29 March had elicited a request for clarification and elaboration of what White had in mind; and it was provided in the very next letter John was to receive, and as the first paragraph:

Selborne, April 29, 1774.

Dear Brother,

   Out of all my journals I think I might collect matter enough, and such a series of incidents as might pretty well comprehend the natural history of this district, especially as to the ornithological part; and I have moreover half a century of letters on the same subject, most of them very long; all which together (were they thought worthy to be seen) might make up a moderate volume. To these might be added some circumstances of the country, its most curious plants, its few antiquities, all which together might soon be moulded into a work, had I resolution and spirits enough to set about it.[23]

There, in a fully-fledged form, is not only *Selborne* but the 'Antiquities' as well. In time (the three years from January/February 1771 to March/April 1774), White had moved from the idea of an *annus* to a prospectus for 'a moderate volume', but in attitude the demands and complexities of moulding 'a work' had replaced the simple chronological collocations that were to have made an *annus*. Stimulated by contact with Pennant, Sheffield, Skinner, Lightfoot and others, cajoled by Barrington into the formalities of his monographs on the

137

swallows, reassured by success at the Royal Society, as well as fortified by the companionship (albeit mostly by letter) of his brother, White had arrived at an insight and strength of purpose that prompted not only the prospectus for a work but also the momentum to proceed. Within 18 months he was arranging to choose an artist to illustrate the work, and Mulso was expecting an announcement of publication in the newspapers. As we know, however, these were premature signals: another twelve or so years were to pass before White was to pen the last lines of his volume. They are years that carry their own story.

Distractions: 1775–1788

To label the delay between prospectus and completion of *Selborne* as a period of distractions is a partial judgement and one that derives from the perspective of posterity. Yet, although White may have been sceptical of Mulso's opinion that what he was engaged upon would 'immortalize [his] Place of Abode as well as [himself]',[24] there is sufficient evidence to suggest that the task was not unappealing,[25] and it would be wrong to think that the decade and more prior to publication were years of self-doubt, indecision or severe slackening of interest. On the contrary, the time shows White richly engaged: not only in maintaining a well-established pattern of living and record-keeping, but also in expanding and improving his estate, in fulfilling increased family responsibilities and in developing new interests. The delay in publication, real though it was, derived (at least initially) from a deliberate, one might say over-conscientious, wish to proceed to an all-inclusive publication.

The crucial decisions occurred in 1775 and 1776. Fully intending to proceed, White first sought guidance from people he could trust (his brother-in-law, Thomas Barker, for example, and Mulso),[26] and wrote up as additional letters several topics not yet explored in his real correspondence with Pennant and Barrington, the most notable being a detailed account of the much-treasured memory of the 1741 cobweb shower;[27] also he chose an artist to illustrate his work. Much care was taken over this invitation, and it gives a clear indication of White's understanding of his potential audience and of the kind of volume he intended. Friends in London were directed to make recommendations and give reports on the work of favoured artists, and in the early months of 1776, after receiving encouraging reports about Samuel Grimm's style, he inspected Grimm's work himself: he was pleased with what he saw and engaged him; and the artist visited Selborne in July, taking twelve views in 24 days.[28]

Concurrent with his rapid progress on the natural history part of the volume, White had begun—as foreshadowed in the April 1774 proposal—to gather data about the 'few antiquities' of the parish. Report about early steps in this direction appears in a letter written in London during February 1776 to Thomas Barker at Lyndon. After asking for any further suggestions concerning aspects of natural history that he should consider, he continued:

> *I have employed the keeper of Domes-day Book to transcribe all relating to Selborne; and am to pay 4d. per line:* besides I have applyed for a transcript of all relating to the Priory in Magd[alen] Coll. archives.[29]

Inspection of the archives at Magdalen was necessary on account of Selborne Priory: after subsisting for 'about two hundred and fifty-four years', it had been suppressed in 1486 and passed to Magdalen,[30] who maintained the estates as a source of revenue. In the event, however, White discovered that he had launched into more than he expected. The first substantial information came from a new acquaintance, Dr Chandler, who ascertained that the parish had sustained not only a medieval priory, but also a preceptory (a community run by the Knights Templars); inconveniently for White, the statutes of the college forbade the archives being removed from Oxford, and for further information he had to attend in person. Chandler had told him that the archives contained much 'information that [had] never been pryed into, but [had] slumbered within the college walls ever since they were founded', and it seems that White pursued his researches eagerly—but not until October 1777. After the visit, he reported to his brother at Blackburn that his discoveries about Selborne had been very great and that with Chandler's assistance he had 'examined 366 parchments'.[31] Digesting data of this extent took time, and there was more to follow, namely, the registers of the bishops of Winchester; and so, on the occasion of the first exploration of this new material (August 1778), White was somewhat downhearted. Parts of the archive were intriguing, even amusing, but, although the place the antiquities were to hold in his work had already been determined, White admitted to some doubt about his progress:

> Should I ever [he wrote to a nephew in 1778] be able to finish my work respecting this my native place, the old deeds and charters &c. will furnish a large appendix.[32]

Assistance was, however, at hand. He had obtained permission for the registers to be brought from Winchester to Selborne, and there, in each of the summers of 1778, 1779 and 1780, he worked on them under the guidance of Chandler.[33]

With such help available, one might surmise that White could rapidly have brought his writing to a conclusion—and published. That, at any rate, was Mulso's view. He considered White's work on the 'Antiquities' an unnecessary embellishment and, writing late in spring 1782, reproved him thus:

> Another Winter is pass'd without your *Essays*. I have no more to say than that You are a timorous, provoking Man: You defraud Yourself of a great Credit in the World: as to your laboring at your Antiquities, it is mal-apropos; the World does not care for such rough work now. Your Porch will be bigger than your House; and You will clap a Gothic Front upon a Plan of Palladio, I mean this, if You *labour too much* at it.[34]

In truth, White was in no danger from labouring *'too much'* at the antiquities of the parish. Although his new interest had brought (by 1782) several new friends[35] and new activities (to both of which he devoted considerable attention), and engaging and distracting as such pursuits were,[36] they did not engulf him as much as other developments—that is, the extension of his property and the growth of responsibilities as head of the family. Noticeably, both developments run alongside the demands made by the antiquities. Wakes will be considered first.

Earlier, during discussion of the relationship between the garden at Wakes and the countryside beyond (especially the hanger), it will have been noticed that, although the main period of activity took place around 1760, there were several later extensions. The most important of these occupied White in the mid and late 1770s, just at the time he was beginning work on the history of the parish: these were the planning and construction of a new hermitage (1776), the purchase of farmland between Wakes and the hanger (1777), and the building of the Bostal (1780).[37] White himself was not involved actively in any of the manual labour and his brother Thomas furnished much of the finance,[38] but the family had no estate manager and White himself had oversight of operations. The account of work on the Bostal should be seen as typical:

> Finding that Larby alone would never finish his job, I hired a whole band of myrmidons, and set them to work . . . they have made great dispatch, and have but half a day's work to come, which has been delayed by the rains. They ran through the upper part a day sooner than I expected, because as we advanced the soil grew shallower; but then we have been obliged to widen and raise all Larby's first attempt, because his path was so narrow, hollow, and clayey that it soon grew

> dirty and would have been impassable. . . . In our progress we found
> many pyrites in the clay as round as a ball, and some large *Cornua
> ammonis* in the chalk.[39]

Once finished (and the whole path was a quarter of a mile in length),
White enjoyed the result, particularly the views back to the village
which, interrupted by the tall beeches, he found 'romantic'. It was the
last substantial addition he made to the prospects available to himself
and to his friends, and it became well used as a path between Wakes
and the home of relatives in the neighbouring village of Newton
Valence.[40]

Alongside these changes to the local landscape, White launched
into the only substantial addition made in his lifetime to the structure
of Wakes itself.[41] The idea of some kind of extension to the house
appears to have originated as a consequence of entertaining brother
John and his wife during the early months of 1773. Evidently, with
guests staying for a long period, conditions proved somewhat
cramped, and only a few months after his visitors left White wrote to
John:

> I thought a fortnight ago that I was going to build a chamber full speed.
> I had bespoke a mason . . . and Jack was to have been comptroller
> general of my majesty's works; but just as I was going to lay in all
> materials my mason sent me word he had got another job, and could
> not do mine 'til after harvest.[42]

Harvest came and went, and over the next winter his plan changed.
Instead of a strictly functional 'chamber', he began to 'talk of building
a parlor'. Mulso was enthusiastic and sent White some enticing
advice:

> Pray, when you build, let it be a 'Drawing Room up Stairs, that
> you may look on the Hanger; Let it be higher than the present, & let it
> be sashed.—Monstrous! why this will be a great Expence!—True,
> therefore take two Years instead of One to do it in. As You want to
> decoy your family after You to make Selborne a Place of Residence, as
> well as to enjoy it during your own Life, e'en do it in a tempting way.[43]

White was hesitant ('I have fears about the trouble; besides green
walls will not be habitable till the second year') and the scheme
appears to have been shelved. Observation and enjoyment of his
brother's extensions at Fyfield, however, renewed his appetite
('Building is very infectious and catching: I am so pleased with
[Henry's] new parlour, that I want to go home and build one'), and
plans were made for the addition of a similar room at Wakes. Some
delay ensued, but building began in 1777 and, interestingly, many of

the important stages were entered in the *Journal*: these embraced the size of the room, the building of the walls and completion of the chimney, plastering, various stages of drying and, eventually, papering. This last took place in June 1780 and in the following month (on 11 July) is the note: 'Finished my great parlor, by hanging curtains, & fixing the looking-glass'.[44] Within a short while he was receiving visitors, and his niece's approval: 'my uncle's new room', she wrote to her brother, 'which is quite finished in every respect and looks very handsome . . . has a very pleasing effect, and . . . is I think one of the pleasantest rooms I ever was in'.[45] Mulso's 'tempting' advice had been prophetic and White must have been relieved at the successful outcome.

The work involved in building the parlour, the adornments to the vistas within and beyond the Wakes estate, the research associated with the antiquities, not to mention the routine duties at Oriel and in the parish,[46] were not the sum total of the distractions that delayed publication. As important as any of these were the pleasures and sadness associated with family matters. By 1780, White was sixty years old; although he was a bachelor, he was head of a large family; and the family home was at Selborne. The responsibilities that flowed from these simple facts were considerable.

From inspection of *Kalendar* and *Journal* the reader might think that mention of visitors, brief data about journeying to London, Oxford and elsewhere, and matters of domestic economy are obtrusive, not just for the purist (for whom the garden alone should figure in the *Kalendar*, and nature in the *Journal*), but for White also. There is some truth in such a view, particularly if the records of the early years are considered and if the private correspondence of the later years is ignored. But the cycle of the years will have its way, and we have to be prepared for some changes in the lifestyle and predilections of our subject. If a representative portrait of White as a young man in the early 1740s saw him with gun in his hand, a spaniel at his heel, intent—through the stubble—on partridge, but mesmerised by the lacy beauty of a shower of cobwebs; and if one saw him in middle age inspecting his melon bed, or listening carefully to the churr of a fern-owl while the framework of the Hermitage perceptibly vibrated; then by the early 1780s a typical scene would show him recording in the *Journal* details of the latest weather and the behaviour of his newly-acquired tortoise,[47] or penning to a niece or nephew advice regarding English poetry (and enclosing a copy of his own latest verses), or supplying news about current events in the village. Moreover, he frequently and generously extended invitations to his correspondents to visit Selborne, whether in summer 'in all its glory, in its full foliage, and verdure', or in winter 'shorn of its tresses, and much in dishabille', for

> we have still pleasant foot-paths, wild views, and chearful neighbours [and] I will give you some roast-beef, plum-pudding, and other Xstmass-cheer.[48]

Indeed, so great was White's pleasure at living the role of paterfamilias that it becomes necessary to seek within those same family connections the explanation for his eventual conquest of distraction, and the appearance of *Selborne*. The varied aspects of domestic life that White so enjoyed in the 1770s and 1780s provide simultaneously both a source of distraction and the very catalyst he needed to spur him to literary activity. Matters concerning several members of the younger generation, and one member of White's own generation, are crucial.

As a man who had no intimate personal attachments but who had made his home in the village with which his family had been associated ever since his grandfather had arrived there as vicar in 1681, it was perhaps inevitable that by 1781 the rest of the family looked to White as a figure of continuity and to Selborne as a retreat from the anxieties of an increasingly-threatened world.[49] A glimpse of this has already been seen in the way White had oversight of the schooling of Gibraltar Jack, and later, although there is little detailed documentation, he undertook something similar for Richard White,[50] a son of brother Benjamin. More obvious is the delight White gained from the company of the younger generation, especially as they passed adolescence and entered upon their life's design. This is particularly true in the case of Samuel Barker (son of White's sister Anne and Thomas Barker of Lyndon), with whom he maintained a regular correspondence for twenty years,[51] and of Mary White (the eldest daughter of his brother Thomas).[52] Without the letters White exchanged with these two relatives, our knowledge would be the poorer in several respects. Writing to members of a younger generation is a very different art from maintaining correspondence with a Pennant or Barrington, and different also from writing to relatives in his own age-group. This is especially true in the letters to both Samuel and Mary. Of all his many nephews, Samuel is the one who shared most of White's interests — in natural history, in travel, in literature; their entire correspondence reveals much about White's current researches, about English poetry, about the general climate of his mind. Somewhat in contrast, Mary White developed an aptitude for antiquarian studies and on several occasions she provided White with useful references and quotations in connection with his researches in that field; she also discharged valuable domestic tasks (such as arranging for supplies of tea and salt-fish to be sent to Selborne), and throughout the 1780s was recipient of White's most inconsequential letters in which he gave news of village events in an informal and humorous fashion.

White's links with the younger generation, many of whom regularly visited Selborne and the neighbouring village of Newton Valance where another nephew Edmund became vicar in 1784,[53] are most characteristically seen through reference to the summers of 1782 and 1783. During 1782 White received visits from several relatives, including two nieces from Rutland who played 'elegantly on the harpsichord . . . [and] entertained us day after day with very lovely lessons from Niccolai, Giodani, and several other modern masters, in a very agreeable manner';[54] and in the following year the *Journal* records in the period June to October a total of 13 nieces and nephews staying at Selborne.[55] The pleasure White obtained from witnessing the delights offered by the locality to these younger members of the family was intense, and as well as enjoying their music and company his Muse—so long silent—flowed once again into poetry and he composed several sets of verses 'for the use of [his] nephews'.[56] The subjects, as before, were pastoral (autumn crocus, scenes of harvest); and they provided, if any were needed, a firm reminder that the lasting truths of life are to be found, not in any human creation, but in the scenes of the natural world.

The distractions of the 1780s were not provided just by the younger generation, neither did the traffic through the Selborne lanes belong only to them. Much as he delighted in life at Selborne, White also travelled—to London, to Oxford, to Fyfield, and to friends nearer at hand—and spent (to take 1783 as typical) very nearly two months away from home. Visiting of this frequency demanded it be reciprocated and, in this same year (1783) that 13 young people stayed at Selborne, so did seven adults, two of whom were relatives, the others being Mr and Mrs Richardson (friends from Bramshott, a village the other side of the forest),[57] John Mulso (from Winchester), Henry Becke (botanist, fellow of Oriel, and a future professor of modern history) and Ralph Churton (who was to become Archdeacon of St David's and who arrived shortly after Christmas to stay for a fortnight).

More notable than any of these various visitors was the addition, in summer 1781, of a remarkable presence to his household. During the late 1770s the health of his brother John had seriously declined, and he had died at Blackburn in November 1780. Gibraltar Jack and his mother then stayed in Lancashire (for Jack to complete his medical training in Manchester), and travelled south the following year.[58]

1. White's extended garden, taken from Samuel H. Grimm, 'North East view of SELBORNE, from the SHORT LYTHE' (*Selborne*, 1789).

2. The Hermitage (App. B, item 7) near the Zig-zag, taken from Samuel H. Grimm, '—where the Hermit hangs his straw-clad cell' (*Selborne*, 1789).

Sept.r 30. The men are weeding the garden, which
is very much over-run with groundsel.
Oct.r 1. A very cold, blustering day. Began fires.
Began gathering the white apples, & golden pippins.
Earthed-up the Celeri, some rows to top. Used ye
first Endive: it is too small to blanch well.
Planted a row of Burnet-plants brought from
ye Sussex-downs.   The cater-pillars have been pick'd
off the savoys several times: those that have not
used that precaution have lost every plant.
The cucumbers, & kidney-beans are cut-down with
the Cold.   The ashes, & maples in some places
look yellow.   The wood-lark sings, & the wood-
pidgeon cooes in ye Hanger. John took his bees.
3. Vast showers with frequent claps of thunder.
Discovered the Enchanter's nightshade, (Circæa) it
grows in great plenty in the hollow lanes.
4. Gather'd in my baking-pears, about three bushels.
The woodruffe, when a little dryed, has a most fragrant
smell.
5. Examined the wild black Hellebore, (Helleboraster
niger flore albo) an uncommon plant in general, but
very common in Selborne wood. Discovered the nipple-
wort (Lampsana.) Vast heavy showers, with a tempestuous wind.

3. White autograph from the *Garden-Kalendar*, fol.
198r, 30 September–5 October 1765 (by permission
of the British Library).

38.

June 25. London-pride, Geum folio subro-tundo majori, pistillo floris rubro, blows:

Common nettle, Urtica racemifera major peren-nis, blows.

Moss-Provence rose blows.

Narrow-leafed bulbous Iris flowers, Xiphion.

Garden-Larkspur, Delphinium, blows.

The horned beetle, called by the French Cerf volant, appears: scarabᵉ max: platyceros, taurus nonnullis.

Crow-foot Geranium, Geranium Batrachoides flore cœruleo, in bloom. The blossom is very large, & shewy.

26. White Ladies-bedstraw, molluginis valgatioris varietas minor, blows.

27. Agrimony, agrimonia, flowers.

Wheat, triticum, blows.

Meadow-sweet, Ulmaria, blows.

28. Ivy-leaved sowthistle, or wild lettuce, Lactuca sylvestris murorum flore luteo, flowers.

Nipple-wort, lampsana, begins to blow.

Devil in a bush, nigella, flowers.

Great purple snapdragon, Antirrhinum purpureum majus, blows on walls, &c.

Garden & wood strawberries are ripe.

4. White autograph, the *Calendar of Flora* for 25–28 June 1766, taken from *A Nature Calendar* (1911), ed. W. M. Webb.

14. April 14. Shepherd's needle, or Venus's comb, scandix semine rostrato, blows.
Lamb's lettuce, vale: verianella, budds for bloom.
Winter-cress, or rocket, eruca lutea, seu barbarea, budds for bloom.
Hen-bane, hyoscyamus, emerges.
Dog's violet, viola martia inodora, blows.
Young rooks in their nests.
15. The vine, vitis, budds.
Polyanth-Narcissus blows.
Young figs appear.
17. Nightingale, luscinia, returns, & sings.
Bloody wall flowers blow.
Privet-leaved lilac, or persian Jasmine, Syringa ligustri folio, buds for bloom.
Raspberry, Rubus idæus spinosus, leafes.
19. Wood pease, or heath-pease, Orobus sylvaticus foliis oblongis glabris, blossoms.
The titlark, alauda pratorum, sings.
20. Jack by the hedge, or sauce alone, Hesperis allium redolens, blows.
Branched whitlow-grass, paronychia ramosa hirsuta, blows. a mistake.
Dyer's weed, luteola, comes into leaf.

5. White autograph from the *Calendar of Flora* (1766), showing (i) calligraphic variation to record the prognostic arrival of the nightingale (17 April), and (ii) acknowledgement of error (20 April), taken from *A Nature Calendar* (1911), ed. W. M. Webb.

Year / Place / Soil		Therm	Barom	Wind	Inches of Rain or Sn. Size of Hail-st.	Weather.	Trees first in leaf. — Fungi first appear.	Plants first in flower: Mosses vegetate.	Birds and Insects first appear, or disappear.	Observations with regard to fish, and other animals.	Miscellaneous Observations, and Memorandums.
Selborne.											
June 18.	Sunday. 8 / 12 / 4 / 8	71.	29 6/10	NW. N.		Cloudless. hot sun with air. sweet even.					Wheat looks well. Phallus impudicus stinks.
19.	Monday. 8 / 12 / 4 / 8	62½	29 6/10	NE. N. NE.		dark. dark &. still. sweet even.					Great honey-dew. Began cutting my meadow-grass. Grass very short.
20.	Tuesday. 8 / 12 / 4 / 8	71 3/4	29 6/10½	W. NW.		great dew. bright. showers about. sultry. sweet even.		Agrimonia eupatoria.			Barley shoots into ear. Meadow-grass very short indeed.
21.	Wednes. 8 / 12 / 4 / 8	71.	29 10½	W. NW. W.		fine dew. sun. fresh air. bright cool gale.			Hay makes at a vast rate.		Rasps begin to ripen. Vast crops of plums, currants & goose-berries. ✗ A house-martin which laid in an old nest: hatches.
22.	Thurs. 8 / 12 / 4 / 8	73.	29 7/10	W. W.		grey. sun & sultry with brisk wind. clouds about.					Pines begin to ripen at Hartley. I have not seen the great species of bat this summer. Ricked meadow-hay in delicate order
23.	Friday. 8 / 12 / 4 / 8	65.	29 10½	W. SW. W.		sun. hot. dark & windy. dark & spitting.					about half a crop: it is very short & fine. The œstrus curvicauda the insect which lays its nits on
24.	Saturday. 8 / 12 / 4 / 8	61½	29 3/10	W. W. NW.		bright. dark. soft rain for hours. grey & mild & still.					the flanks & legs of horses appears in the forest: it seems to abound most in moist moorish places, tho' sometimes seen...

Teals breed in Wolmer-forest: jack-snipes breed there also no doubt, since they are to [be] found there the summer thro': — A person assures me, that Mr. Meymot, an old clergyman at [a] chappel in Sussex, kept a cuckow in a cage three or four years; & that he had seen it several times winter & summer. It made a little jarring noise, but never cryed cuckow: It might perhaps as hens. He did not remember how it subsisted.

6. White autograph, *The Naturalist's Journal* for 18–24 June 1775, with marginal annotation in the hand of Daines Barrington (by permission of the British Library).

7a and 7b. White letter, 30 August 1770, to John White at Gibraltar, offering advice to his brother on the writing of natural history (bMS 731/113, by permission of the Ralph Waldo Emerson Memorial Association and of the Houghton Library, Harvard University). *(Continued overleaf)*

Dear Sir,

I have nothing to add to my former account, but that your son continues in good health, and has begun his Greek Gramm[ar]; in which he goes on very well. My Wife and Daughter join me in respectful Compliments, and I am ever, Dear Sir, Your faithful hble Serv[t]

Aug[t] 20:1770

R Willis.

Dear Brother, M[r] Willis & Jack wrote more than a month ago: that letter, I trust, is come to hand before now: You need not fear but that I shall keep all y[r] papers & letters respecting Nat: history. If I might advise, concerning your work in hand, I should think you should address it to y[r] friend, Govern[r] Cornwallis. A short but respectful dedication void of all flattery would be very becoming. You might then add a short preface, mentioning y[e] inducements for entering on such a work; & remark that Nat: hist: when considered in a physico-theological light is no ways foreign to the pursuits of a clergyman. Yet if stile be clean, easy, & correct: & in that respect few men can acquit themselves better than you can. Your work may open with the civil history of Gibraltar from y[e] earliest times: in this matter I shall app[y] to M[r] Toot Greek profess[r] & the Hon: M[r] Barrington for anecdotes. You must mix with y[e] nat: descriptions now & then a dissertation, as y[e] journey to the Emperour of Morocco, &c. Publish a journal of one entire year with its coincidences: & the other incidents naturally occurring escaped you in that year: you may fairly insert them from a preceding or following Journal. M[r] Stillingfleet, you see, has opposed the Calendar of Upsal & Stratton. Now Sweden is about as far N: of England as Gibraltar is S: therefore (if you approve of it) I can furnish you with a diary from Selborne for y[e] year 1769: & which means those climates will be exactly compared so as to afford some information to a curious naturalist. My Journal will be very full & particular: so that both to-gether will afford perhaps 180 pages. Y[r] Journal will be wanting with respect to it's Flora: however mention the blowing

of as many plants as you can. Where are your smelts caught? How are the montagoes, or Spanish hams cured; or if the hogs from whence they are taken fed with vipers? I have sent your birds & fishes to M[r] Pennant; but have not received his answer yet. As far as I have yet been able to examine, Hirundo hyberna is entirely new: we shall see what he will say. I have given one specimen to my friend M[r] of Worcester Coll: who is an excellent naturalist: we think at present that it is a perfect nondescript. I have compared it with the bank martin: but it differs widely in shape & colour. Y[r] Motacilla No: 7 was the stone chatter: they were not by any means rare: I thought at first they might have been cold-finches; but they were not: 19.24 were wagtails: Muscicapa 80 was Edwards's grey fly-catcher: No: 9 was the common regulus non cristatus: Y[r] 41 is a variety. Look for the Scarabaeus pilularius, or tumble-turd. This is the insect that in Aesop's fables dropped balls of dung into Jupiter's lap among the eggs of the Eagle, & occasioned that God's shaking out the eggs. Wherever nat: circumstances explain any fable or passage, lay hold of them, & apply them. You did well in vindicating the fable of the fox & grapes from the imputation of improbability. I have got I think a fine motto out of Homer's Odyssey for the climate of Andalusia. Yet more, I see, they are very rare. M[r] Sheffield had never seen one: M[r] Drake in his wonderful collection of insects in E[ngland] had but one. You will find probably, if you examine, they breed in y[r] inundation. Among y[r] fishes were perca marina, tho' you mentioned no such genus, &c. summer declines, or rather when it is convenient, send me more birds, the mellba & oropendulo, &c. & a barbary sheep. When do y[e] land-rails appear & disappear? The aurora boreal: that figured so much with you on the 24 of Oct[r] made the same appearance here. This implies that those meteors are higher in the atmosphere than I was aware of, if they play with us also all day in the season. Can't you make a dissertation on the currents of the streights of Gibraltar. Y[r] red-billed alca is the common puffin. You have the credit of finding the winter retreat of that genus. Compare y[r] tridactyla with the common gull. I want to make it a new species: aboard or are they only a casual variety. Have you no

The single sheet, folded to provide four pages, was begun by Gibraltar Jack (White's nephew) on 18 August, added to by R. Willis (Jack's schoolmaster at Holibourne, near Alton) on 20 August, and completed by White.

Jack pursued his career, first at Alton, later at Salisbury, but Mrs White stayed at Wakes and accepted White's invitation to take charge of his household and run the domestic affairs. For the first time since the death in 1755 of his grandmother, one senses that with the running of Wakes in secure and capable hands White himself felt able to turn his attention more freely to literary matters.

That Wakes from this time on became something of a summer paradise for the ever-growing White family has been suggested already, but there was another consequence. Scrutiny of the *Journal* indicates that Mrs John White made more demands on the garden, for the purposes of culinary preparation, than White had been used to providing; and he appreciated this extension of his already strong interest. Entries in the *Journal* for 1785 are representative:

> 4 July: Gathered several pounds of cherries to preserve; they are very fine.
> 10 July: Preserved cherries, & currans; & made curran-jelly.
> 22 July: Made black curran-jelly, & rasp: jam.
> 25 July: My Nep: Edmd White sends me some fine wall-nuts for pickling.

In other years equally evocative entries appear about vegetable produce, for example:

> 1784 (15 Sept.): The autumn-sown spinage turns-out a fine crop . . . We draw it for use.
> 1786 (25 June): Cauliflowers, Coss-lettuce, marrow-fat pease, carrots, summer-cabbage, & small beans in great profusion, & perfection. Cherries begin to come in: artichokes for supper.
> 1787 (1 Nov.): Split-out the great Monk's-rhubarb plant into 7, or 8 heads, & planted them in a bed that they may produce stalks for tarts in the spring.[59]

This enthusiasm for the produce of the garden (and of the hanger: there are regular entries in the *Journal* for truffles),[60] instead of distracting from the work necessary to complete *Selborne*, seems actually to have stimulated him and renewed his appetite for the task. With Mulso, Churton, Loveday (the Caversham antiquary), and many relatives warmly in support, there was no lack of encouragement, and as the 1780s advanced so did the writing.

Stronger than any of these influences was Mrs John White herself. She was a woman of determination and great purpose to whom authorship was not unknown. At Gibraltar, and later at Blackburn, she had witnessed the strenuous demands made on her husband by his Gibraltar fauna, the projected 'Fauna Calpensis'.[61] Faced now with a brother-in-law who was ever civil to visitors and villagers alike and who delighted in laying aside his pen to attend (at any time of the day or night)[62] the fascinations of the natural scene, she must surely have added in her own way to the support White could now depend upon. At any rate, the long-planned work came to an end and in January 1788 White wrote:

> I have been very busy of late, and have at length put my last hand to my Nat. Hist. and Antiquities of this parish. However, I am still employed in making an Index — an occupation full as entertaining as that of darning of stockings, tho' by no means so advantageous to society. My work will be well got up, with a good type and on good paper, and will be embellished with several engravings. It has been in the press some time, and is to come out in the spring.[63]

In fact, publication took place in the following autumn (the engravings are dated 1 November 1788),[64] and the long journey of composition was over. The judgement of the public was awaited.

1789 and After

Copies of the book, which was published in three different bindings, were sent to friends and relatives and early acknowledgements were received by White from Henry (who noted in his Journal: 'Hamper from London containing ye Natural History of Selborne, presented by ye Author. A very elegant 4to with splendid Engravings & curious invest^s [investigations].'),[65] and from Mulso (who was generous in praise and let White know that the Chapter at Winchester Cathedral had determined that a copy should be purchased for the Cathedral Library). In 1789 reviews appeared in *The Gentleman's Magazine* and in *The Topographer*. This latter was of especial acuity:

> The book is not a compilation from former publications, but the result of many years' attentive observations to nature itself, which are told not only with the precision of a philosopher, but with that happy selection of circumstances, which mark the *poet*. Throughout therefore not only the understanding is informed, but the imagination is touched.[66]

More interesting than these immediate outcomes, however, was the effect of completing the book on White's own writings: this is seen in the *Journal*, in a projected essay, and in correspondence.

Although the pace of entries in the *Journal* never slackened over the years of preparation for publication, when the work was finished several perceptible changes occurred. With more time to devote to writing up his observations, extended entries on particular interests become more prevalent. Some of these entries reflect a broadening of definition of what is natural history (for example, there are accounts of the building in London of a huge conservatory for the Queen of Naples, of a new form of fire-escape, of meeting a survivor of the siege at Gibraltar); others (for example, reflections on the sloughing of a snake skin, and remarks on the roosting habits of poultry and other table birds) are of a form, style and content that, had they been written earlier, would readily have admitted them to the published volume.

Two further characteristics of the *Journal* entries from 1789 to 1793 are noticeable. Over the years, White had always made notes on interesting phenomena that reliable witnesses had reported, and on less common findings that were brought to him for comment. Now, the status of author seems to have considerably enhanced his local reputation and such entries occur more prominently: especially in 1790, when he received on separate occasions a specimen of a Manx shearwater, *Puffinus puffinus*, and a bird that he pronounced to be a mule, that is 'a spurious or hen bird, bred between a cock pheasant and some domestic fowl'.[67] Yet, if the community publicly acknowledged White as an expert in the field of natural history, he himself seems to have spent more time than he had earlier with the writers of the past who had incorporated observation of the natural scene into poetry. Admittedly, the *Journal* throughout its completion is punctuated by quotations from many authors; but in these final years, although use is made of the works of natural historians (Ray, Derham, Willughby, Linnaeus and Scopoli are all consulted), it is quotations from the poets that catch the eye. Moreover, whereas in the past such quotations, when they occasionally appeared, were used didactically, now they enter the text as confirmatory wisdom; as if, after all the detail that had been amassed in his various journals and after all the observations that he had made, White had come to acknowledge that the higher truths of the poet embraced satisfactions closest to the needs of men in their search to understand the benignities of the natural world.[68]

It was not only in the *Journal* that White maintained a contribution to enquiries about natural phenomena. In summer 1789 he was brought eggs of a nightjar and shortly after wrote in correspondence:

> I have just found out that the country people have a notion that the *Fern-owl*, or *Eve-jarr*, which they also call a *Puckeridge*, is very injurious to weanling calves by inflicting, as it strikes them, the fatal distemper known to cow-leeches by the name of puckeridge. Thus does this harmless, ill-fated bird fall under a double imputation, which it by no means deserves,—in Italy, of sucking the teats of goats, where it is

called *Caprimulgus*; and with us of communicating a deadly disorder to cattle.[69]

The injustice of these charges against a bird that ranked in his regard as affectionately as any swallow prompted the desire to disabuse the world of its erroneous understandings, and the idea was conceived of writing up his observations of many years in a paper for the Royal Society. Support for the worth of an up-to-date account came from the dearth of detailed knowledge of the bird as shown by several standard works (for example, neither Willughby nor Ray seemed to know it was a summer migrant), and White collected material as widely as he could, including information from Cheshire supplied by Ralph Churton. Unhappily, when the paper—which, White asserted, would 'make a fierce appearance with a quotation from *Aristotle*, & another from *Pliny*'[70]—was nearly ready (during the winter 1792/3), he discovered that his advocate at the Royal Society, Daines Barrington, no longer had influence to determine the admission of papers, and White's final piece of sustained writing about the natural world was lost to posterity. Opportunity to seek another advocate, or to publish elsewhere, did not present itself, and we have been deprived of further evidence of White's skill in his chosen field.

However, we are not totally bereft of White's work in these last years. In addition to the *Journal*, there is a joyous series of letters that White exchanged from 1790 until his death with an admirer of *Selborne*. Robert Marsham had written to White in July 1790 to convey the 'pleasure and information' he had obtained from reading White's book. Prompted by topics discussed by White, he commented in detail on migration, on the meteorological records he kept in Norfolk (his home county), on the natural calendar and on his especial interest—the planting and tending of trees.[71] Topics such as these alone would have been sufficient to recommend this new correspondent, but when he discovered Marsham to have been a friend of his own early mentor, Stephen Hales (for whom he had discharged the duties of curate at Faringdon), and that it was at Marsham's house that the author of the *Calendar of Flora* (1755), which had set him on to the studies of his own *Flora* (1766), had kept his records, then his delight was unbounded. 'O, that I had known you forty years ago!' he exclaimed in a letter of December 1791. It was a sentiment Marsham reciprocated. A bond initiated by common interests was confirmed by friendship, and the correspondence flourished. In these last years, it provides the clearest evidence of White's continued enthusiasms and of his capacity to extend those enthusiasms in new ways. Marsham acknowledged White to be the expert in ornithology; as for White, he was guided to an enhanced understanding of arboriculture. Many of the letters discuss the girth and age of trees, and also their height; and data are exchanged about oak and ash, chestnut and sycamore, elm and beech. Indeed, in the very last composition of his life, a letter to Marsham of 15 June 1793, the central topic is an oak in the forest adjacent to Selborne that Marsham granted was 'the largest in this Island': White wished his friend to know that at that very time drawings of 'this extraordinary tree' were being prepared for subsequent engraving and publication.[72] The mottoes for his own long-enduring work on *The Natural History and Antiquities of Selborne* had been fully vindicated.[73] His native soil, rough and 'mountainous' as it might be, had been demonstrated to be richly fertile: he was to die on 26 June 1793 with this knowledge secure.

CHAPTER TWELVE

The Natural History of Selborne

Awareness of the wholesome strengths and enduring fertilities of the Selborne countryside must not be seen as an experience unique to White alone, or even to those contemporary with White: so long as what he wrote is available, and we are prepared to read it, the experience can also be shared. But we are faced with a dilemma, for there are two opportunities open to us: either we can read the raw data, namely the *Kalendar, Flora, Journal* and many of the private letters, or we can read the finished *Selborne*. In a sense the dilemma is unreal: all the texts referred to are readily accessible and we can choose to pursue both opportunities; and that is what the present volume invites us to do. Yet the experience offered by the one kind of text is not the same as that offered by the other. The materials already discussed, whether journals or letters, were working documents written under the pressures of activity and observations. *Selborne* is different, and if it is to be read aright some adjustment of expectation is needed.

At first sight, White's published volume, with its 'large appendix' devoted to the antiquities of the parish,[1] is an elegant account by means of letters addressed to Thomas Pennant and Daines Barrington of its chosen subject, the natural history of Selborne, the introductory letters affording an appropriate historical and geographical context for what follows. Inspection of the contents shows the care taken to distinguish matter based on observation from matter that is speculative, since the writer is persuasively concerned to be 'as sure of the certainty of [his] facts as a man can be of any transaction whatsoever'.[2] This is not all: everything mentioned in the text seems real. Places, whether villages, towns or counties, the works of other naturalists (many of which are cited), and the phenomena themselves can all be verified by the reader.[3] In circumstances such as these, where the writer's declared integrity is allied to a factuality of subject matter and reference, it must seem preposterous to propose an approach to White's text that is different from that pursued in connection with earlier documents. The attempt must, however, be made since *Selborne* is not an accurate, scrupulously documented and coherent scientific record—which is what one might reasonably expect as one moves to read it after the *Journal*—but a literary record, dotted with lacunae, illiberal in logic, provocatively disposed to solve all problems by an appeal to Providence; in sum, a volume subject not to the perceptible rhythms and occasions of the natural world, but beset by the vagaries of correspondence and the chances of art.

To substantiate strictures as severe as these must await the general availability of the copy text from which *Selborne* was set by the printer,[4] but sufficient evidence for the claim to be seriously addressed already exists. It is set out in Appendix G as an exploration of four examples of White's procedures in the preparation of *Selborne*, but I here provide a fifth that is especially illuminating.

Letter 29 in the *Selborne* sequence addressed to Pennant is dated 12 May 1770. It falls into two parts. The first, which comments on the cold spring and the late arrival of migrants, closely follows a real communication to Pennant of the same date. The second part, prepared solely for publication, provides several anecdotes about the pairing of birds, comments on the dislike cats have for water and their fondness for fish, and concludes with:

> Quadrupeds that prey on fish are amphibious: such is the otter, which by nature is so well formed for diving, that it makes great havock among the inhabitants of the waters. Not supposing that we had any of those beasts in our shallow brooks, I was much pleased to see a male otter brought to me, weighing twenty-one pounds, that had been shot on the bank of our stream below the *Priory*, where the rivulet divides the parish of *Selborne* from *Harteley-wood*.

The passage shows several characteristics of White's writing: interest in the physical form of a creature, record of its weight and place of discovery, awareness of the status of hunter and hunted, delight in an addition to the local fauna. More importantly, if we search White's writings the passage can be compared with the following:

> Two young men killed a large *male otter*, weighing 21 pounds, on the bank of our rivulet, below Priory longmead, on the Hartley-wood side, where the two parishes are divided by the stream. This is the first of the kind ever remembered to have been found in this parish.

Undoubtedly, this second passage is the source for the inclusion of the item in the *Selborne* letter to Pennant dated 1770. Yet, and this is the point to be seized, the shooting of the otter occurred some 14 years later, for the passage is taken from the *Journal* and is entered at 3 October 1784.

To find a level of inconsistency as substantial as this last must, taken with the examples given in the appendix, prompt grave doubts about the reliability of many aspects of the whole of *Selborne*. Although such doubts would be understandable, it would be wrong on that basis to challenge the reliability of the entire text. To reject White's published volume because of a few (very few) inconsistencies and errors would be as inappropriate a response as to accept it unreservedly because of an apparent absence of 'error'. Nevertheless, that there are aberrations alert us to the fact that we need to find some way of reading the volume that falls neither into rejection nor into blind acceptance. Before that is attempted, however, the scale of the changes from raw observations to published letters (of the kind noted during discussion of the report concerning the otter) must be established, and some remarks made on White's use of the letter form itself.

Sequencing and Arrangement

How, then, did White proceed to edit his letters and journals? What was his purpose? What kind of work did he have in mind? To what tradition, if any, did he see his work contributing? Unfortunately, except for *Selborne* itself, there are few data available to help in answering these questions. What I termed earlier the 'prospectus' (of 29 April 1774: see Chapter 11, page 137) tells us only that the content might comprise 'incidents . . . [of] natural history of [the] district', together with 'some circumstances of the country, its most curious plants, its few antiquities', and no later reference to the real format of what was intended exists. Comment must rely, therefore, on evidence gathered from the final work and its underlying materials.

The primary text is the manuscript of *Selborne* that was sent to the printer. It is an intriguing document, from which White's several stages of editing his materials can be reconstructed; and it demonstrates in a remarkable way two aspects of *Selborne*, an understanding of which is crucial for reading the volume in the spirit in which it was prepared. These are the shaping of individual letters, and the shaping of the whole.

Some awareness of what White accomplished when editing his correspondence has been gained from considering the letter that included report of the otter. That showed that a letter with a given date (Pennant 29) was made up from part of a real letter of the same date, from some specially written anecdotes and from a version of a *Journal* entry of a much later date. Two further (and different) examples will now be explored. The first involves two letters dated 14 September 1770 and 29 October the same year addressed to Pennant (Pennant 31 and 32 respectively). That they both begin with a location, 'Selborne', and a precise date suggests they are closely derived from real letters. This is so: genuine letters for the same dates (one of which carries a seal) addressed to the same correspondent exist in White's hand; but they are scarcely recognisable as the letters that appear in *Selborne*. Naturally, some differences must be expected: to turn, in a period of decorum, real correspondence however formal into a publishable volume required 'restraint and impersonalization',[5] particularly since the focus of the intended volume was parochial rather than personal. One effect of this is the difficulty there would be if, from *Selborne* alone, it were required to establish, say, that the author had one brother who was a natural history bookseller, another who planned a history of Hampshire and a third who had completed a natural history of Gibraltar, or that the author, a fellow of an Oxford college, regularly performed clerical duty and could count among his relatives several dozen nephews and nieces. The results of this de-personalising process are vividly illustrated in the two letters under discussion. The real letters, considered together, include mention of visitors at Selborne, comment on his correspondent's travels and thanks for the gift of a book, reference to his brother Thomas's absence from London and his own visit to Ringmer ('where I stayed three weeks, & from whence I wrote a long letter to Mr: Barrington'), and extended discussion of Gibraltar specimens together with report that his brother's inquiries in that territory were much restricted since the death of his horse.

With the deletion of so much—and the real letter of 29 October almost entirely disappears since its central subject is Gibraltar specimens—White was left with very little usable material. Nevertheless, he must have thought that, because letters about natural history were actually written on the given dates, something from them should be included in his book. As a consequence the real 14 September letter provided not only the content for the whole of the *Selborne* letter of that date, but also most of what appears in the 29 October letter. More interestingly, the Selborne manuscript reveals that this disposition of material had been accomplished within a few months of White's announcement in March-April 1774 to his brother at Blackburn of his intention to prepare a natural history; and that the distribution made at that time remained substantially unchanged right through to the published work some 14 years later.

If White could edit real letters by the addition of extended observations taken from the *Journal*, and distribute a single letter across letters of different dates, albeit addressed to the same correspondent, he could also change his addresses. The *Selborne* letters of 15 January 1770 and 19 February the same year are the third and fourth letters in the sequence directed to Barrington. Both letters are derived from a single real letter of 15 January 1770 addressed to the same correspondent. This real letter, however, was very long; and since the exchange of letters with Barrington had begun only recently (in the previous year), it contained very few, in fact scarcely any, homely personal references of the kind mentioned above in the letters to Pennant. This meant that, when he came to edit the letter, White found much more material than could be fairly assigned to a single publishable item.

That letters to be published in a book might have a limitation on their length may seem unusual, but throughout *Selborne* White held in the main to a very clear notion that a natural history letter was a literary form that required each letter to be devoted to a single theme or topic. This being so, when editing his 15 January letter he did three things. One theme was birds that sing—that became the substance of Barrington 3 of the same date, 15 January; another theme was the behaviour of the cuckoo—that became Barrington 4 of 19 February, a date recalling a real but unused, because homely, document to Barrington. But this still left a substantial portion of the letter unused: a passage concerning the food in winter of birds such as hedge sparrows, robins, titmice. White's answer was to address it, undated, to Pennant! Quite why he made this decision is unclear, but a possible answer lies in its position in *Selborne* as Pennant 41.

Correspondence with Pennant had started in 1767, but by the early 1770s Pennant had transferred his interest from obtaining information about Selborne fauna to seeking information from John White about fauna at Gibraltar, and in the belief that there was little more White could contribute the exchange of correspondence ceased by the mid-1770s. The letter preceding Pennant 41, however, is dated 2 September 1774, just at the time, it has been argued, that White began seriously to fabricate his work; the following letter, Pennant 42, is dated 9 March 1775. This suggests a procedural sequence. Beginning in spring 1774, attention was devoted first to editing the letters to Pennant, a process that by late autumn that year had reached those of autumn 1770, Pennant 31 and Pennant 32.[6] Following this, White then completed work on the remaining Pennant letters—to September 1774, Pennant 40; he then turned to his Barrington correspondence. The first two Barrington letters, of 1769, were entire, but as he came to that of 15 January he decided, knowing his exchange of letters with Pennant was almost over, to assign part of it to the Pennant sequence, in the hope perhaps that he might be able to maintain both sequences.

Confirmation that something like this was intended is given in the next letter in *Selborne* addressed to Pennant (Pennant 42): although it has the precise date of 9 March 1775, it is derived from an earlier, real letter of 19 March 1772. Only two further letters occur in *Selborne* in the Pennant sequence. They both suggest composition in 1780, a year referred to in each. White's intuition had been right: his regular correspondence, at least with Pennant and soon with Barrington, was over, and the task, not of editing that correspondence into natural history letters, but of deciding how best to follow it up lay before him. Evidence of this remains in *Selborne*. Although the correspondence with Pennant and Barrington had been substantial in quantity, it was severely restricted in range. Birds were what White knew best and what both his correspondents had encouraged him to write about. Admittedly, there had been some further explorations (for example, into reptiles and fish) but now much more was required. If the volume was to be comprehensive, as its title was to imply, then additional materials would have to be found.

It was this stage of thinking that accounts for the letters that occur in the Barrington sequence introducing topics as various as gypsies, the making of rushlights, the account of the boy for whom bees 'were his food, his amusement, his sole object', and village superstitions.[7] They occur in *Selborne*, it can be noticed, almost immediately after the most focused discursive letters White ever achieved—those on the house martin, swallow, sand martin and swift, which he had prepared first for the Royal Society in 1773-4. Several of the topics, suggestions for which he had sought from relatives, coincide with letters he was writing in the mid-1770s to his nephew, Samuel Barker, and it is very clear that what in *Selborne* appears at this time as an indecisiveness of direction reflects White's own uncertainty. It also seems likely that, as well as using ideas suggested by family correspondence,[8] some consideration was given to what in his own period (and later) had become the archetypal natural history. This was *The Natural History of Oxford-shire* (1677) by Robert Plot, to which there is reference in *Selborne*. It was from this volume that White obtained the idea of enquiries contributing, as he remarked in a letter to Pennant (Pennant 31), 'to an universal natural history', and also accepted guidance about the investigation of specific topics such as the echoes in the parish (Barrington 38).

But more important to any evaluation of the content of *Selborne* must be what stands as introduction and conclusion. It has been noticed before that the natural history portion of the volume ends with Letter 66 addressed to Barrington and that the remainder, 'The Antiquities', is an appendix. Yet the shape of the main portion was decided very late; as was the decision to leave the appendix unaddressed, probably because, although it was presented in the form of letters, they were composed solely for publication. The Selborne manuscript shows that what now begins the sequence of letters to Pennant (Pennant 1-9) was written as literary, that is putative, letters to Barrington (for example, Pennant 1 and Pennant 2 were originally marked as a single letter, Barrington 53; and Pennant 3 and Pennant 4 were marked as Barrington 54). In arriving at these decisions, which were of great significance to the character of the final work, White took advice from relatives and it is evident that some of the finishing touches to the volume were determined by them rather than by the author.[9] To an extent greater than has been usually suggested, the whole work might legitimately be considered a collaborative volume: its beginnings can be found in real correspondence with Pennant and Barrington; confidence to publish derived from grappling with the consignments of specimens from Gibralter and with the drily academic writing of brother John; and now members of the family were assisting in editing his manuscript in ways he was pleased to endorse. White himself would not, I think, resist such thinking.

Use of the Letter Form

With an understanding of what might be termed the 'fabrication' of *Selborne*, it is now possible to consider the local format of the whole: that is, the use throughout of the letter as the unit of composition. Authors of natural history works prior to White's had organised their material in one of several formal ways,[10] and his choice of an informal mode, the letter, was against expectations if the outcome was to make a serious contribution to the discipline of science. And yet, as noted above, it was through correspondence that White had first discovered that the world valued what he had to say, and it must have seemed very natural for him to have decided that the immediacy and authenticity of what he wrote would be best served by the retention of the letter form. Further, sanction for the use of letters could be adduced from two, seemingly contrary, directions: the scientific, and the literary.

Study of the history of scientific inquiry during the late seventeenth and early eighteenth centuries shows that written communications, customarily in the form of letters, were the primary means by which scientists reported results and exchanged seinformation.[11] The most pervasive expression of this exchange was the *Philosophical Transactions* of the Royal Society, direct experience of which White had obtained through addressing his letters on the hirundines to Barrington, who had then arranged for them to be read at meetings of the society. He was also familiar with a volume of John Ray's correspondence which had been edited by Derham and published posthumously in 1718. This volume, *Philosophical Letters*, is mentioned several times in *Selborne* and White drew from it more than has been previously noticed, particularly in the editing of his real letters. As Derham remarked in his preface, after he had decided which letters to publish:

> I thought it necessary to leave out all that might be of little use, such as private Business, Complements, &c. except now and then a Clause, that may be of use to Mr. Ray's, or some other learned Man's Character, or that may shew their learned Projects, or give some Account of their Labours.

It would be difficult to find a simpler statement of White's dealings with his own letters. In retaining the epistolary format he allied himself immediately with the best of a tradition that had furthered the cause of science for many decades before his own letters were published. But the form was not exclusive to the world of science; it flourished also in literary circles and is sometimes seen as the most characteristic mode of the century.[12] In the works of Samuel Richardson (discussed on several occasions by Mulso in his letters to White) epistolary fiction had achieved a penetration and subtlety that has scarcely been bettered, and Defoe had shown how to incorporate in a single volume real letters alongside fabricated letters.[13]

To a modern reader unaccustomed to the conventions of a past age, this absence of a clear differentiation between what is factual and what is fictional may cause difficulty. Taken at face value, letters may include factual report or pursue the fanciful, but a natural history must surely be written on a basis of observation. As has been shown, *Selborne* is written in letters, some of which are real, some made up; letters that purport to be real are shown to be fabricated, and the truly real may be spliced to the speculative; further, the whole work seems to satisfy neither the criteria expected in the reporting of scientific investigation, nor the freedoms of imaginative creation. In itself this would seem to discredit the volume, since it was published at a time when the division of knowledge into arts and sciences was becoming noticeably more apparent.[14] In effect, the volume's form, as well as its content, should be regarded as an important statement of what White accomplished.

The world that *Selborne* entered in 1788–9 was a world bemused by uncertainty, threatened by revolution, riven by rancour. The order of things was changing dend *Selborne* must have seemed no more than a dying ember of a world that had passed, the world into which White had been born nearly seventy years before. If we wish to understand *Selborne* properly, it is to the first decades of the century that we must look; for they secured for White enormous advantages. After a childhood of pastoral security, and education at home and school, he had graduated at a university that alongside its support for the traditional orthodoxies of politics and religion and its keen interest in the classical world had made a leading contribution to scientific pursuit; moreover, in Plot it had nurtured the originator of county histories.[15] To these tendencies there had been added the strongly practical and scientific bias of Hales, and the opportunity, over the years of endeavour recorded in the *Kalendar*, to fashion the cultured (and cultivated) resources of garden and hanger. Then, fired by the Stillingfleet programme, White mastered the latest of classifications for the natural world, the Linnaean system. He focused his inquiries on the restricted fields of, firstly, his own parish and, secondly, by proxy, his brother's environment at the Rock; and, through acquaintance and exchange of information with leading naturalists, stood poised ready to enter the scientific circles of the day and make an orthodox contribution.

But it was not to be—and for reasons of character. The flamboyant coteries of the Royal Society were not for him, either intellectually or by interest and inclination.[16] Abstract speculation, dry prolixity or grand theorising were not his *métier*. Nothing could interpose between his accumulated knowledge of the variety and wonder to be discovered in his native place and the opportunity to engage the world with an account of its life and conversation, and he turned away. Instead, he chose, as he put it in his 'Advertisement' to *Selborne*, to be a 'stationary' man, intent to induce 'a more ready attention to the wonders of the Creation . : . [to contribute to an] enlargement of the boundaries of historical and topographical knowledge . . . [and to throw] some small light upon ancient customs and manners'.

To accomplish these differing purposes White was forced to select. From the days, weeks and years of accumulated, incremental records locked in letters, journals and in his own memories, he created a pattern, a figure of meaning, in a volume called *Selborne*. Integral to the figure was the decision to mediate his understanding of the world by means of letters that, whether fabricated or real, invited the reader to participate in the author's own journey of local exploration, to share the very motions of his mind, its doubts alongside its certainties, its astonishment in discovery, and, most important of all, the satisfaction obtained from investigative activity itself.[17] A listing of the volume's contribution to distinct discoveries in its field of natural history is meagre,[18] and any engagement with the serious scientific issues of the day negligible.[19] To ask whether *Selborne* is a scientific or a literary record is a fragile and inconsequential enquiry. At its centre lies an exploration that embraces and transcends both. Because White centred his interest and methodology on the life and conversation of animals, it can be claimed that he wrote the original textbook in behavioural ecology.[20] But of far greater purpose is the communicative value of the volume. For White, the final justification and consolation was religious and social: purposeful questioning into the natural productions of his parish

> by keeping the body and mind employed [had] under Providence contributed to much health and cheerfulness of spirits, even to old age

for his pursuits had

> led him to the knowledge of a circle of gentlemen whose intelligent communications . . . afforded him much pleasing information . . .

White knew that 'the poetry of earth is never dead', and as he moved through the beeches on the hanger or glimpsed the first annual swoop of swallow or martin he felt assured of 'the dearest freshness deep down things'.[21] Essentially, *Selborne* is a work of communal adoration, seeking an answering echo in each and every reader to the mystery of the eternal birth and rebirth of the natural scene, and to the distinctively English countryside, temperate, sweet.

Notes

CHAPTER ONE

1. *Selborne*, Letter 1 to Pennant. Later items of this kind will normally be referenced by name of correspondent followed by the number of the letter, e.g. Pennant 1. The full title of White's volume is *The Natural History and Antiquities of Selborne, in the County of Southampton: with Engravings, and an Appendix* (London, 1789), an acknowledgement of authorship appearing only at the end of the prefatory 'Advertisement' in the form 'Selborne, / January 1st, 1788. GIL. WHITE'. Copies of the work were available in late autumn 1788, since there was (and is) a convention which permits books published in the final weeks of one year to carry the date of the year following.

2. There have been four bridges over the Thames at Hampton Court. The present bridge (of 1933) was preceded by one built in 1865, this last replacing one of 1778. White must be referring to repairs made to the original bridge of Chinoiserie style erected in 1753. The 'Toy', which White mentions in his letter, was an inn adjacent to the Trophy Gates of the Palace; at one time it was used as a barracks and 'Victualling House' by the Parliamentary Army. It was demolished in 1857.

3. The account is of interest in several ways. Although White points to the longevity of this particular nesting site, modern observation shows that a single pair of ravens prepare to nest in up to five or six (even seven) different sites. As for species of tree, considerable variation exists: in Germany beech is preferred, in Sweden pine; today in England species of conifer are most used (Coombs, 1978). Of equal interest is White's neutrality of tone (possibly approval): in the face of so much traditional superstition about the raven as a harbinger of death, and distaste for it as a carrion-feeder, it might be thought that villagers and boys would approach their task with caution, even apprehension. But these were rational times, and work on the tree would be rewarded, White's conclusion centring on the strength of the providential maternal instinct, rather than on any regret at loss of either oak or raven.

4. The 1703 tempest, the subject of Defoe's *The Storm*, began 'on Friday 26 November and lasted until Wednesday 1 December [and is described as] probably the worst ever experienced in England . . . It was estimated that 8000 people lost their lives in the floods' (Brazell, 1968: 11).

5. One of these bequests led White to keep accounts headed 'Lands in Hawkley called Collier's . . . being my Grandfather White's Charity for Schooling' from 1760 until his death. The income from the land was usually about £1 per annum (although this amount trebled from 1780 onwards), and White used it to pay for the schooling by a local dame of a number of children (Gilbert White Account Book (folio, 55 leaves), ff. 4v–8r, Gilbert White Museum, Selborne).

6. Robert Marsham (1708–97) was previously unknown to White, but he had written a letter of congratulation on read-

ing *Selborne*; he was a great lover of the natural scene, specialising in knowledge about the growth of trees. Letters between Marsham and White are given in Bell (1877), **2**: 243–303).

7. This couplet is from the translation (1753) by Joseph Warton of *Georgics*, Bk II, 525–6; White, in his letter to Marsham (Bell, 1877, **2**: 249), uses the Latin—'Et dubitant homines serere, atque impendere curas?' (*Georgics*, Bk II, 433).

8. Edward Young, *Satire* VII in *Love of Fame* (1728), lines 21–4. The earlier verses from Pope are from 'Windsor-Forest' (1713), lines 397–400, 369–70.

9. The title-page of *Selborne* gives the parish as situated in the County of Southampton, since such was its administrative area until reorganisation in the 1950s; White's use of 'Hampshire' refers to an ancient geographical territory comprising roughly the present county of Hampshire together with the Isle of Wight.

10. A good example of this is his record of Howe's relief of Gibraltar in 1782. The *Journal* notes Howe's departure from Spithead (11 Sept.), arrival at the Straits (11 Oct.), completion of the relief (19 Oct.), skirmish with the combined French and Spanish fleet (20 Oct.), and return to Portsmouth (14 Nov.), with the observation, such was White's patriotism: 'He was absent just nine weeks.' Later, he bought a copy of a contemporary account of the siege (which lasted four years, and is generally thought of as one of the most memorable in history).

11. Bell (1877, **2**: 284): the troops were German, from Hesse, and were later deployed in America.

12. For the text of this account, together with a detailed commentary, see Foster (1986c).

13. White's letters to his brother when the latter was chaplain at Gibraltar are given in Foster (1985a).

14. For example, the accounts list volumes as varied as James Ray, *A Compleat History of the Rebellion . . . in 1745*; Robert Watson, *The History of the Reign of Philip the Second, King of Spain*; Edward Gibbon, *The History of the Decline and Fall of the Roman Empire*; Charles F. Sheridan, *A History of the late Revolution in Sweden*; and John Drinkwater, *A History of the late Siege of Gibraltar.*

15. The threat of invasion was felt keenly almost throughout the second half of the century, one of the more exciting periods being in 1759 when someone as urbane as Horace Walpole was induced to write on 22 June: 'Well! they tell us in good earnest that we are to be invaded . . . I hear of mighty preparations. Of one thing I am sure; they [the French] missed the moment when eight thousand men might have carried off England, and set it down in the gardens of Versailles'; and, in the same year, on 8 July: 'I don't write to tell you that the French are *not* landed at Deal, as was believed yesterday. An officer arrived post in the middle of the night who saw them disembark. The King was called up; my Lord Ligonier buckled on his armour. Nothing else was talked of in the streets'—and yet, as the style reveals, Mulso's apprehension was foolish. Quotations from Walpole (1960: 299–300, 305).

16. See Mulso (1907: 280) for this and the preceding quotation. Voltaire's *Candide* (1759) was an exercise in ridiculing eighteenth-century self-satisfaction; somewhat typically, John Mulso (1721–91), an Oxford contemporary and lifelong correspondent, prefers to ignore the satire and read the withdrawal into the privacy of one's garden literally. The letter in which he proffers White this advice was written in 1778, and may be found in Holt-White (1907).

17. In *Selborne* (Barrington 37), White wrote that 'Potatoes have prevailed in this little district . . . within these twenty years only; and are much esteemed here now by the poor, who would scarce have ventured to taste them in the last reign [that of George II (1727–60)].' See Chapter 4.

18. The detail is much worse than my generalisations suggest: for example, there was no electoral contest at all in Shropshire between 1722 and 1831, and in '1780 only two English county seats actually polled' (Porter, 1982: 127).

19. Urban population was not to surpass that of the country until 1851.

20. I have in mind here the woodpigeon, *Columba palumbus*, a bird White terms the ring dove; but damage in winter to turnips from stock doves, *C. oenas*, was also considerable. Writing of these latter, White noted in a *Selborne* letter dated 1780 (Pennant 44): 'The food of these numberless emigrants [they 'traversed the air . . . in strings, reaching for a mile together'] was beech-mast and some acorns; and particularly barley, which they collected in the stubbles. But of late years, since the vast increase of turnips, that vegetable has furnished a great part of their support . . . and the holes they pick in these roots greatly damage the crop.' For the gardener the depredations of a more domestic creature had to be faced, a note in the *Journal* for 1 Sept. 1785 reading: 'Dogs [ate] the goose-berries when they became ripe; & now they devour the plums as they fall: last year they tore the apricots off the trees.'

21. Some wells (on the Downs) were up to 350 feet deep, so White discovered (*Journal*, at 18 Dec. 1773), but his own well and others at Selborne were only 63 feet; and yet, in the severe drought of autumn 1781, the well at Wakes did not fail.

22. Although White gives a detailed account of winter lights prepared from rushes, the practice was by no means universal (*Selborne*, Barrington 26). In the next century, William Cobbett (1822) was also advising a similar mode of lighting.

23. '. . . sometimes I have seen one tree on a carriage, which they call there [Sussex] a tug, drawn by two and twenty oxen, and even then, 'tis carry'd so little a way, and then thrown down, and left for other tugs to take up and carry on, that sometimes 'tis two or three year before it gets to Chatham; for if once the rains come in, it stirs no more that year, and sometimes a whole summer is not dry enough to make the roads passable' (Defoe, 1724–7, Letter II). For improvements in the course of the century, see Moir (1964).

24. Conditions at Selborne seem to have been exceptional, since, although White avers 'We abound with poor', he continues 'many . . . are sober and industrious, and live comfortably in good stone or brick cottages, which are glazed, and have chambers above stairs: mud buildings we have none' (*Selborne*, Pennant 5).

25. Porter (1982: 31, 35).

26. The examples, and others, may be found in Keith Thomas (1983). White himself saw his garden as the more important when birds threatened his produce: 'We have shot 31 black-birds, & saved our goose-berries' (*Journal*, 8 Aug. 1781), and 'Pd Will Dewey for 8 Doz: of young sparrows' (*Kalendar*, 6 July 1765), are two examples of many.

27. The earliest London hospital of the century was the Westminster, founded in 1720, and provincial hospitals at York, Salisbury and Cambridge were actually a little earlier; but nationally there were 14 or so other hospitals established between 1730 and 1750, and many others in the second half of the century. A detailed list is given in G. S. Rousseau (1978).

28. See *Selborne*, Pennant 34; the earlier quotation, concerning grassland, is from Barrington 40, White's allusion to 'two blades of grass . . .' being a reference to the King of Brobdingnag's advice to Gulliver: '. . . whoever could make two Ears of Corn, or two Blades of Grass to grow upon a Spot or Ground where only one grew before . . . would deserve better of Mankind, and do more essential Service to his Country, than the whole Race of Politicians put together' (Swift, *Gulliver's Travels* (1726), Bk II, Chap. 7). White's use of 'commonwealth's man' must not be thought to have any overtones of puritanism or anti-monarchical tendencies: he was a loyal royalist.

29. An enduring and accessible literary contribution to the dispute was Swift's *Battle of the Books* (1704).

30. John Ray, *The Wisdom of God manifested in the Works of the Creation* (12th edn, 1750: 104, 199).

31. The most important example of this is his discovery that swifts mate on the wing (*Selborne*, Barrington 21).

CHAPTER TWO

1. The date was 13 February. It is worth noting that dates in the first half of the century need careful treatment. England adopted the Gregorian calendar in 1752 (by which time the Continent was eleven days ahead), the change being made in September: that is, 2 September 1752 (Old Style of calendar) was followed directly by 14 September 1752 (New Style). At the same time, opportunity was taken to make the next year begin on 1 January, and not—as would have occurred under the Old Style—on Lady Day, 25 March. A remnant of Old Style remains with us in the dating of financial years. All dates given in this volume are New Style.

2. For a comprehensive history of Wakes, see Meirion-Jones (1983).

3. Details of the family are given in the Genealogical Chart, Appendix A.

4. White to his niece, Mary, May 1784 (Holt-White, 1901, **2**: 29).

5. *Selborne*, Antiquities, Letter 6.

6. 'Cleaned my well by drawing out about 100 buckets of muddy water . . . Nothing had been done to this well for about 40 years. The man at bottom in the cleaning brought up several marbles & taws [the fancy, streaked marble with which one shoots] that we had thrown down when children' (*Journal*, 1 Oct. 1781).

7. See respectively Holt-White (1901, **1**: 31–3); *Selborne*, Antiquities, Letter 26; *Journal*, 10 Mar. 1793. Additional material, and discussion, appears in Scott (1946: 24–9).

8. Quotations from this work, in many ways the representative poem of the century, were frequently used by White, especially in the *Journal*: see, for example, 23 Aug. 1774, 26 Oct. 1783, 14 Sept. 1791.

9. Rye (1970: 50, 52) describes him as 'interestingly passive or negative', and refers to him as 'harmless', 'odd . . . strange'. Mulso (1907) is the best source for information about John White's character, for it is there we can read of his musical powers (he played the harpsichord), charity, and 'christian Temper'; extant autograph material (items 28 and 29, in the Holt-White Collection currently at the Gilbert White Museum, Selborne) is also useful.

10. Bell (1877, **2**: 259); White says the estate was seven acres, laid out by his father in walks and hedges. Detail of John White's professional training is given in the records of the Middle Temple: he was admitted to the Inn on 26 Nov. 1707 and called to the Bar on 15 May 1713; from 4 Mar. 1711 he occupied No. 1 Elm Court, and held a life interest in the chamber until his death on 30 Sept. 1758, at which time it 'Fell to the House'.

11. Mulso (1907: 19).

12. Bell (1877, **2**: 32).

13. Both notes occur during what was to prove John White's final months; they are dated, respectively, 29 Jan. and 18 Mar. 1758 (*Kalendar*, ff. 54r, 57v).

14. For full accounts of White's ancestors, both paternal and maternal, see Holt-White (1901) and Scott (1946). The forthright figure of Gilbert White senior is shown well in his family portrait, which can be seen at the Gilbert White Museum, Selborne.

15. Mulso (1907: 95).

16. But he seems to have enjoyed the 'Tangles of Neaera's Hair', and may have wished to be married (Mulso, 1907: 7, 99–100). That he once translated an ode by Horace, Bk 3.26, that can bear the title 'Love Renounced and Resumed' (Marshall, 1911) is perhaps also relevant (Mulso, 1907: 16).

17. *Selborne*, Barrington 8 and Pennant 6.

18. The only financial accounts to be published (except for some brief entries in the *Kalendar* relating to garden dung, and a few references in Holt-White (1901)) are for daily expenses Mar. 1752–Mar. 1755 (Bell, 1877). Some extant autograph accounts are now at Selborne, at the Gilbert White Museum. Discussion of those for 1740–5 is in Emden (1948: 129–32).

19. 'Dinner' in the eighteenth century was taken at 3.00 p.m. or thereabouts; White paid one shilling for such a meal.

20. Bell (1877, **2**: 326). There are amusing later entries for expenses in connection with the dog 'Fairey', the payment of three shillings and sixpence for 'Advertising, & crying Fairey' and subsequent payment of seven shillings as a 'Reward to Bernard Bailey at Swallowfield [to where] she was strolled'. The total cost of the 'strolling' was more than this since a further entry notes the payment of one shilling to the 'News-man for bringing her to Selborn'. Swallowfield is about thirty miles north of Selborne on the road to Oxford, and as White had only just arrived home for Christmas it seems that Fairey took against the excursion into the country.

21. Mulso (1907: 293).

22. *Selborne*, Pennant 7. In fact, at this time, the 1780s, the sporting scene in terms of 'prey' was beginning to change: deer, because of enclosures, developments in road systems, and the demands on forest timber, were in retreat and the era of the pheasant was just beginning. This last occurred largely because of advances in the construction of guns which permitted the 'flying' pheasant to be shot in the air, traditional game had been the hare, and the ground-hugging partridge. A useful compendium of attitudes to game, which included rabbits and pigeon, is *A New Treatise on the Laws for Preservation of the Game* (1766), and Hopkins (1985) offers a lively account of attitudes in the next century.

23. *Kalendar*, 20 Feb. 1765.

24. White himself was to report such events in the *Journal*: see entries for 15 Aug. 1772 (relating to Nore Hill); for 16 Sept. 1775 (relating to events on 8 Sept. at Oxford and Bath); and for 14 Sept. 1777 (relating to Manchester).

25. Yet he would not have sanctioned the extreme view as enunciated by Soame Jenyns: 'If we look downwards, we see innumerable species of inferior beings, whose happiness and lives are dependent on his [Man's] will; we see him cloathed by their spoils, and fed by their miseries and destruction, inslaving some, tormenting others, and murdering millions for his luxury or diversion' — quoted by Willey (1950: 52). For exposition of the hierarchies in nature, the classic text is Lovejoy (1936).

26. *Selborne*, Barrington 23.

27. Yet memory of the horn room at Wilton House (Salisbury) is noted (*Selborne*, Pennant 30); and Mulso (1907: 4, 9) mentions Burghley House (Stamford) and Windsor, both, apparently, being places where White enjoyed the pictures. In 1783, much later than these visits, White bought a copy of Horace Walpole's *Anecdotes of Painting*, and it seems possible that he had a latent interest in pictures throughout his life; certainly, at the time he employed S. H. Grimm to take some of the Selborne views (to accompany his own text), he took a keen interest in the artist's work (Bell, 1877, **2**: 54, 55).

28. For Whitwell, and Thorney, see Bell (1877, **2**: 164, 260); Ringmer was the home of his aunt, Mrs Rebecca Snooke, whom he visited regularly until her death in 1780, and Mulso (1907: 7, 8, 9) shows that White was there for several weeks as early as 1745.

29. A reliable source for this kind of data is Holt-White (1901); for data on White's contemporaries, Foster (1888) can be consulted.

30. Because of legislation at the Restoration which forbade dissenting clergymen from teaching, many of the most forceful thinkers of the age established their own academies and contributed to the education of men as varied in intellectual and political stance as Isaac Watts, Samuel Wesley, Defoe, Bolingbroke, Bishop Butler, and Priestley, Hazlitt, Malthus, and Godwin. But our understanding of the realities of education at Oxford in mid-century is beginning to change, and the view that the university was moribund (attributable in part to Gibbon) is being seriously questioned: see especially Sutherland (1973) and Markham (1984: 66).

31. Mulso (1907), and White in a brief memoir published in *The Gentleman's Magazine* (1781: 11–12), are important sources for the biographer of Collins.

32. Mulso (1907: 4).

33. The poem is present in some edi-

tions of *Selborne* (for example, Bell, 1877); it was first included in the Mitford edition of 1813—see Appendix D. For the Horatian ode, see Note 16 above.

34. White's 'Expenses in the Small-pox' are given in Holt-White (1901, **1**: 49, 52).

35. This was Tom Mander, an Oriel contemporary of White's, who was especially interested in physics. Mulso (1901: 16), writing at a time White was in contact with Mander, asks: 'How does Tom Mander's System of Physics go on? Is He Master of ye weight & ye Power? Has He settled Sr Isaac's [Newton] & Grimadi's Dispute of ye Refrangibility or Dispersion of Rays? will He venture down in a diving Bell, or is He yet as distress'd as a Cat in an Air Pump? You may give my Love to Him, if his apparatus does not forbid your Approach.' Playful as the tone of this is, we know that in later life also White was interested in things mechanical (for instance *Journal*, 14 June 1789, where he describes a new fire-escape).

36. See Chap. 6.

37. Mulso (1907: 45, 46).

CHAPTER THREE

1. Barker kept diaries throughout most of his life, but this early volume appears to be no longer extant: citation of the entries quoted is from Holt-White (1901, **1**: 31).

2. Bell (1877, **2**: 164, 165); Mulso (1907: 14); *Selborne*, Pennant 23; and White accounts at Oriel College, Oxford.

3. Although Benjamin was certainly working at Whiston's shop from 1745 onwards, his partnership seems to have run from at least 1752 to 1765; he was in business on his own account from 1766 until his retirement in 1792. See Mulso (1907: 11, 57); Nichols (1812, **3**: 668–71).

4. These quotations are from letters of White dated 8 Aug. and 27 Aug. 1783 (Holt-White, 1901, **2**: 102, 107). Barker lived to a ripe age, not dying until 29 Dec. 1809, in his 88th year. It was reported of him that he was 'a remarkable instance of abstemiousness, having totally refrained from animal food; not through prejudice of any kind, or from an idea that such a regimen was conducive to longevity . . . but from a peculiarity of constitution which discovered itself in his infancy' (Nichols, 1812, **3**: 112).

5. The influence of opportunity and weather is well shown by noting the dates on which annual entries begin. For the first ten years, these are: 1751, 7 Jan.; 1752, 4 Mar.; 1753, 1 Jan.; 1754, 5 Mar.; 1755, 6 Jan.; 1756, 23 Jan.; 1757, 1 Jan.; 1758, 1/2 Jan.; 1759, 19 Jan.; 1760, 17 May (on return from Lyndon).

6. It may also be significant that Barker had already made his first communication to the Royal Society: it was read 14 Dec. 1749, and published in *Philosophical Transactions*, **46**: 248 (see also Letters and Papers II, **6**). Subsequently he was to submit over thirty further papers, most of which also appeared in *Phil. Trans.*

7. Hales' status as a scientist is due primarily to his ascertaining that vegetable sap does not circulate in a fashion analogous to the blood of mammals—see Stephen Hales, *Vegetable Staticks* (1727); but his more homely pursuits, all essentially practical, are what White most admired—see Bell (1877, **2**: 261–3, Letter to Marsham, 25 Feb. 1791).

8. Commenting on an item in the report of the Bishop of Winchester's visitation at Selborne Priory in 1387 that 'the sacramental plate and cloths of the altar . . . were sometimes left in such an uncleanly and disgusting condition as to make the beholders shudder', White notes: 'Strange as this account may appear to modern delicacy, the author [White himself], when first in orders, twice met with similar circumstances attending the sacrament at two churches belonging to two obscure villages. In the first he found the inside of the chalice covered with birds' dung; and in the other the communion-cloth soiled with cabbage and the greasy drippings of a gammon of bacon. The good dame at the great farm-house, who was to furnish the cloth, being a notable

woman, thought it best to save her clean linen, and so sent a foul cloth that had covered her own table for two or three *Sundays* before' (*Selborne*, Antiquities, Letter 14).

9. John White (1727–80) was to have a decisive effect on White's writing of *Selborne*, but at this time his behaviour caused the family some embarrassment. He had been admitted scholar at Corpus Christi in 1746 and graduated in 1749, and was now keeping terms in the hope of a fellowship. His lifestyle was somewhat irregular, and after several contretemps with college authorities he was expelled on 11 July 1750 for being party to a marriage between 'a young Gentleman of good Family, eldest Son to a Baronet, and Heir to a large Estate, under the Age of nineteen, who was just enter'd Gentleman-Commoner' at Corpus Christi and a daughter of the innkeeper at the Lamb, Wallingford, 'a House of no good Character', and for subsequently entertaining the bride and her sisters 'in his room at College' (quotations from the evidence submitted to the Visitor of the college by the President and seven Fellows, 20 Sept. 1750). The Visitor, the Bishop of Winchester, Benjamin Hoadley, after taking evidence and discussing the issue with the Patron of White's uncle Charles (vicar of Bradley), made his 'Determination', 9 Feb. 1751, in favour of the college. The papers of the case are in the archives at Corpus Christi; the college 'Punishment Register', B/5/1/4, is also relevant.

10. Mulso (1907: 50). How serious White's own thoughts about preferment were is not easy to gauge; but Mulso's uncle, Dr Thomas, at that time bishop at Peterborough, had recently refused translation to the see of 'Litchfield & Coventry', and Stephen Hales had just accepted appointment as 'Clerk of ye Closet to ye Princess of Wales, & that, by ye King's own Nomination' (Mulso, 1907: 26, 47). Perhaps, then, White was already thinking that he might aspire to the provostship of Oriel, as in fact he was to do four years later.

11. On this occasion, Mulso playfully reminded White that his letters already provided 'a pretty Collection of [his] Travells about England' and that, when they were complete, they would 'enable young Men to travel with Taste & improve at Home'; as for his health, he comments on White's return: 'I envy You your bold Flights, your Eagle Ranges . . . I am a poor sculking Quail, whose very Love-Song is plaintive' (Mulso, 1907: 53; see also pp. 79, 97).

12. Holt-White (1901, **1**: 70).

13. In anticipation of a visit from White, Mulso had written on 11 Mar.: 'With what Pleasure shall I receive my old Friend, when just freed from the Confinement & Form which Greatness demands? I reckon You will exult & wanton in your Liberty wth that Glee which we used to go down to Merton Walks to observe in the Horses which were turned out to their Spring Grass' (Mulso, 1907: 66).

14. White's 'season' was from 9 July to 30 Aug.: see Bell (1877, **2**: 336). In contrast to the long spa tradition at Bath, the spring at Bristol had been discovered only about fifty years before White's visit and, although termed a hot well, its temperature is modest — about 70°F as compared with temperatures at Bath of up to 120°F. In about 1725, Defoe commented that the water was 'now famous for being a specifick in that otherwise incurable disease the diabetes' (Defoe, 1724–7: Letter VI).

15. Mulso (1907: 70, 98).

16. Bell (1877, **2**: 345).

17. Gibson was soon to become a naval chaplain and to see service at Louisburg and Quebec, whence he sent White a graphic account of Wolfe's success: see Foster (1986c), which enables White's previously unidentified 'chaplain' (from whom he obtained an account of a moose killed in the St Lawrence river, which in turn contributed to his interest in the famed Goodwood moose: see *Selborne*, Pennant 30, 28) to be named.

18. Bell (1877, **2**: 338).

19. This was John Ray's *Methodus Plantarum* (1682), later published in a totally revised form as *Methodus Emendata* (1703): see Raven (1942: 287–94).

20. *Selborne* (1789: iv), Advertisement. Confirmation of White's own satisfaction at his decision is given by Mulso, who wrote on 18 Oct. 1753: 'I am well pleased to hear from Yourself that You are settled for a Time & in a Place to your Liking . . . I envy more your being equal to your present Employ than your former, for health is the first earthly Blessing' (Mulso, 1907: 72).

CHAPTER FOUR

1. The *Kalendar* appears to run from Jan. 1751 until Mar. 1771, since these are the dates respectively of the opening and closing entries. In fact, the records concerned with the annual cycle of activity in the garden are confined to the years 1751–67 inclusive. All other records, that is those dated 1768–71 (and beyond, into 1773), relate to the purchase of dung, the brewing of beer and the bottling of wine, and overlap in time with the first six years' records of the *Naturalist's Journal*.

2. Such a record might be that of a gardener who was a slave of the calendar: regardless of conditions, both regional and local, sweet peas, say, would be planted out on 1 June, shallots set on the shortest day and pulled on the longest, and parsley sown on Good Friday. These practices must now seem rather quaint, touched as they are by superstition; but I have a sense that they were very common amongst gardeners trained in the later decades of the last century and in the first of this.

3. Of the records White kept before this date, the largest number are given in the *Kalendar*; but of equal (if not greater) importance is a document known as the *Calendar of Flora*. This is a record centred on a single year (1766), but because of its name (which is misleading) it is frequently overlooked in consideration of the documentation that underlies *Selborne*; it contains more than is suggested by the botanical focus of its title and represents a crucial stage in White's development (see Chap. 7).

4. The 'Bob' is the bobtail (or bell) of his clerical wig (Mulso, 1907: 72).

5. All the quotations are from the *Kalendar* (see, for example, 1761: 26 Jan.; 9, 10 Feb.; 10, 16–17, 19, 25 Mar.). It is sometimes argued that the earthworks of 1759 (digging the ditch and building the bank) created a construction that was White's 'first' ha-ha, and that the work of 1761 was associated with a second, more durable construction. Certainly, the ditch and bank of 1759 was 'fenced with sharp'ned piles' (*Kalendar*, 31 Mar. 1759) and the 'Grubbing for, & digging' were accomplished the previous autumn (see White's accounts at Selborne), but such details are not conclusive—especially since similar preparation would have been necessary prior to the works of 1761, and nothing of this is mentioned in the *Kalendar* or in the accounts.

6. Thacker (1979: 181).

7. Two of the most famous of the eighteenth-century designers, and founders of the 'landscape' tradition. For biographical data (and a richly illustrated text) see Miles Hadfield, Robert Harling, Leonie Highton, *British Gardeners* (1980). There is now an extensive literature concerned with the development of the garden, much of it focused on the early years of the century. Important scholarly texts include: John Dixon Hunt, *The Figure in the Landscape* (1976); Maynard Mack, *The Garden and the City* (1969); Edward Malins, *English Landscaping and Literature 1660–1840* (1966); Peter Martin, *Pursuing Innocent Pleasures* (1984); and relevant articles are often carried by *Garden History*, the journal of the Garden History Society.

8. An engaging and well-illustrated account of gardens at Oxford in White's time will be found in Batey (1982).

9. For example, White visited Blenheim and Stowe on 5–6 June 1752, and was again at Stowe only a month or two later, on 11–12 Aug: see expenses given in Bell (1877, **2**: 319, 321). The ha-ha at Blenheim was constructed (although not apparently named thus)

before 1712, and that at Stowe not long after (Thacker, 1979: 184).

10. It is recorded, for example, that a quite small estate employed seventy men for two years to carry out Brown's plans, and at Blenheim the cost was in excess of £20,000. To earn this latter figure White would have to have been a proctor at Oxford not for the single year (1752–3), but for over 160 years: see Turner (1985: 60) and Bell (1877, **2**: 333–4).

11. Perhaps this is overstated, but what successes there were derived from practical knowledge, the chemistry of soil and plant not becoming securely established until the second half of the next century with the publication by a German, Julius Sachs, of *Experimental-Physiologie* (1865).

12. I refer here to the frequent mention of scattering seed for forest trees; a typical example, especially in later years, reads: 'I sent a woman up the hill with a peck of beech-mast which she tells me she has scattered all round the down amidst the bushes & brakes, where there were no beeches before. I also ordered Thomas [White's servant and gardener] to sow beech-mast in the hedges all round Baker's hill' (*Journal*, 22 Nov. 1786). The practice was encouraged by Thomas Barker (Bell, 1877, **2**: 165) and was intended mainly to ensure plentiful timber in future years, but the prospective aesthetic value would not have been inconsiderable.

13. And so, I think, would White; and yet, apart from hill, hanger, and stream, and the presence of certain buildings, nearly everything else would seem different. For example, if he returned in March/April, 'Where,' he might ask, 'are the women weeding the corn, and why is there no smoke from the "vast heath-fires" of Wolmer Forest?' (see *Selborne*, Pennant 5 and 7).

14. Mulso (1907: 15, 24).

15. Mulso (1907: 35); a fascinating account 'of the Sacredness and Use of Standing Groves' is given by A. Hunter in his 1801 edition of John Evelyn's *Silva: or a Discourse of Forest-Trees* (1664).

16. That White so termed certain views comes to us from the usage of the young sister of Mulso, the future Hester Chapone. Writing to White in 1749, Mulso quotes from a letter received from his sister, who was staying at Peterborough: the view from her bedroom window, she wrote, '*as White says, enlarges me*'. See Mulso (1907: 24).

17. *Selborne*, Barrington 17.

18. Mulso (1907: 57–8).

19. The historic past of Selborne was more distinguished than its present, and White ensures that his readers are aware of this. For example, Domesday (unusually) mentions its church; Henry III granted a charter for a market; the Knights Templars had a preceptory and other property; and there was a large priory (from the ruins of which White took 'stones of a sandy nature' to use as coping stones for his fruit wall: *Kalendar*, 25 July 1761). White's reference to the several items can be found in *Selborne*, Antiquities, Letters 3, 10, 11; a fine modern account of the priory is Le Faye (1975: 47–71).

20. This was first published in the 1813 edition of *Selborne*, and the text of that version (for the poem was frequently revised) is given in Appendix D. Comment on some aspects of the poem is in a letter from Mulso of 1751 (see Mulso, 1907: 48–51), and an extended version can be seen in Frank Buckland's 1876 edition of *Selborne*.

21. Mulso (1907: 34, 206); in each instance Mulso appeals to places mentioned by the Latin poet, Horace, in his *Odes*—Bk III, 4 and Bk I, 7, respectively. Acherontia is in some ways unfortunately named: it makes no reference to the River Acheron (part of the classic Hades), but is the name of a pastoral paradise of Horace's youth where, after an exhausting excursion, he 'lay down and the legendary doves / Wove me a blanket of the leaves just fallen . . . [and he was able] Tucked in [a] coverlet of bay and myrtle / To sleep on, safe, a babe / And unafraid, watched over by the gods' (Horace, 1978: 147). The 'grove of Tiburnus' is part of Roman Tibur (Italian Tivoli, about 20 miles east of Rome), where Horace wrote many of his odes: see Highet (1957).

22. Virgil's description (considerably longer than what White quotes) of this ideal 'working' garden is in *Georgics*, Bk IV, 125 ff.

23. Pope, *Moral Essays*, Epistle IV (addressed to Richard Boyle, Earl of Burlington), 'Of the Use of Riches', lines 50–6.

24. Mulso (1907: 90, 92).

25. Later, a considerable disadvantage resulting from vistas was discovered: 'Trees will not subsist in sharp currents of air: thus after I had opened a vista in the hedge at the E. corner of Baker's hill, no tree that I could plant would grow in that corner: & since I have opened a view from the bottom of the same field into the mead, the ash that grew in the hedge, & now stands naked on the bastion, is dying by inches, & losing all it's boughs' (*Journal*, 25 Nov. 1775).

26. Some confusion exists about these structures, because commentators tend to think that White used the terms 'hermitage' and 'alcove' interchangeably. I do not think this is so. In a letter to his nephew dated July 1776 (shortly before Grimm came to draw Selborne scenes), White wrote: 'We have built a new Hermitage, a plain cot; but it has none of the fancy and rude ornament that recommended the former to people of taste: this is strong and substantial, and will stand a long while, fire excepted' (Bell, 1877, **2**: 126). It is this same structure that White referred to in a letter of 8 Dec. 1791 to his brother, Benjamin: 'I have just fixed a white cross, & capped the Hermitage with a new coat of thatch, so that to us below it becomes a very picturesque object' (Rylands 1306/33). Both this new Hermitage and the old Hermitage adjacent to the Zigzag can be seen in Grimm's view of Selborne (Plate 1), and should not be confused with White's *Journal* entry of 2 Oct. 1783 which records: 'Erected an alcove on the middle of the bostal'.

That White built structures and cut paths on land that did not belong to him seems to cause unnecessary anxiety. For example, a recent writer comments: 'Gilbert . . . had only the most tenuous connections with Magdalen College [the Col-

lege was Lord of the Manor at Selborne]; yet nowhere in his correspondence or the college records is there any suggestion that he bothered to obtain permission to cut the Bostal, or for that matter build any of his other constructions on the hill' (Mabey, 1986: 181). In the absence of village doctor or schoolmaster, the Whites as clergymen—and resident in the village at either vicarage or Wakes since 1681—were pre-eminent; they would also have been on excellent terms with the President of Magdalen, who would be entertained on the occasion of the annual manorial court. In fact, at the time of the building of the Bostal the President was George Horne, with whom White was 'well acquainted' and about whom he wrote: 'he has often been at my house' (Bell, 1877, **2**: 287). The procedure for 'planning applications' was rather different in White's time from that of today.

27. John Milton, 'Il Penseroso', lines 168–72.

28. Mulso (1907: 220–1); letter of 1770, Jenny then being about twelve years old.

29. Catharine Batty termed her diary for the occasion 'A little Journal of some of the Happiest days I have had in The happy Valley in the year 1763': a full version of the diary is given in Holt-White (1901, **1**: 129–38). From Catharine we also know of the 'Inscription over the Door of the Hermitage', which read:

> May at last my weary age
> Find out the peaceful Hermitage
> The hairy gown & mossy cell
> Where I may sit & rightly spell
> Of ev'ry star that heaven does shew
> And ev'ry herb that sips the dew
> Till old experience do attain
> To somthing like prophetic strain
> (Harvard 731/58).

Much later White 'Nailed-up a Greek, & an Italian inscription on the front of the alcove on ye hanger' (*Journal*, 10 Oct. 1788).

Opportunity for elaborating the siting of inscriptions was considerable, indeed a characteristic of the period, and in Dec. 1785 *The Gentleman's Magazine* published a sequence of three: the first 'for the entrance of a solitary walk, leading to a hermitage; hung upon a tree with a seat under it'; the second 'for the entrance of the hermitage'; the third was 'to be placed within the hermitage'.

30. This did not exhaust the value of a Hermitage, which could also yield scientific data:

> You will credit me, I hope, when I assure you that, as my neighbours were assembled in an hermitage on the side of a steep hill where we drink tea, [a churn-owl, *Caprimulgus europaeus*] came and settled on the cross of that little straw edifice and began to chatter, and continued his note for many minutes: and we were all struck with wonder to find that the organs of that little animal, when put in motion, gave a sensible vibration to the whole building! (*Selborne*, Pennant 22)

31. A modern historian notes: 'A writer in the *Connoisseur* (25 March 1756) speculates facetiously on the revenue which might be raised by a tax on heathen gods in the gardens of the great: the nobleman at his seat, the esquire at the hall-house, the citizen in his country-box, and even the divine at his parsonage, all have their walks and gardens peopled with satyrs, fauns, and dryads. [He then quotes the contemporary critic thus:]

> While infidelity has expunged the Christian theology from our creed, taste has introduced the heathen mythology into our gardens. If a pond is dug, Neptune, at the command of taste, emerges from the bason, and presides in the middle; or if a vista is cut through a grove, it must be terminated by a Flora, or an Apollo' (Sutherland, 1948: 142).

32. For this emphasis I am indebted to a personal communication from Robert Williams, the Italian garden being that at Villa D'Este, Tivoli. The importance to White of the Italian strain, especially in its classical forms (albeit mediated in part through things of the Renaissance and later) can scarcely be overstressed. For an important treatment of the influence of the Italian Renaissance garden on the English imagination, see John Dixon Hunt, *Garden and Grove* (1986); see also, Dustin Griffin, *Regaining Paradise: Milton and the Eighteenth Century* (1986), and John Prest, *The Garden of Eden: The Botanic Garden and the Re-Creation of Paradise* (1981).

33. *Kalendar*, 24 Jan. 1758; the construction of the statue is mentioned by Mulso (1907: 16). The appropriateness of White's choice of deity is confirmed by Langley (1728), who gives lists of figures for different areas of the garden, and by Spence (1747: 114), who remarks that Heracles

> was pointed out by the antient heathens, as their great exemplar of virtue: and indeed, as the idea of virtue with them consisted chiefly in seeking and undergoing fatigues with steadiness and patience, they could scarce have chosen a fitter pattern than Hercules; the course of whose life was almost wholly taken up in going about to seek adventures; and in labouring for the benefit of mankind.

34. Two principal purchases were 'the upper part of Lassam's orchard' (see *Kalendar*, June 1760), and in 1777 'the fields behind my house, that *angulus iste* [quiet retreat] which the family have so long desired' (Bell, 1877, **2**: 62); both purchases were from John Wells. As for indentures, fees of £200–£300 appear to have been common—and could just as easily be lost (Bell, 1877, **2**: 42, 178). Such costs as these last, and large families, explain the vexed question of White's lowly inheritance, which one writer attributes to his father being 'effectively disinherited' by White's grandfather (Mabey, 1986: 76). Patterns of inheritance no doubt varied from family to family, but a key principle for the Whites was equitable distribution: that Gilbert White senior had expended considerable sums on the education of White's father as a barrister was sufficient reason for him not to inherit Wakes *merely because he was the eldest*—there were also four daughters to be provided for. Expression of this principle figures

significantly in John White's own will, which reads in part: 'All the rest of my estate . . . I give . . . for the only use & benefit of my seven younger children in equal shares as near as may be; taking into the account *what has already been advanced by me to any* of ym in yir education or other wise' (Rylands 1306/1).

35. Milton, *Comus*, lines 386–97.

36. Historically, tents are probably more Arabic than Turkish, but their prime function was the same: to provide, in conditions of fierce heat and intense light, bearable shade. A late (Gothic) development of the tent in the form of a stone folly can still be seen at Painshill, Surrey.

37. Bell (1877, **1**: 507); the fragment is a translation of lines from Euripedes, *Danae*.

CHAPTER FIVE

1. These eight years are those that best represent the content of the *Kalendar* as a document of activity in the garden; before 1757 the annual records are incomplete (owing primarily to White's frequent absence from Selborne), and after 1764 other concerns (especially observation of flora and fauna) intrude.

2. Some of these items may be unfamiliar. 'Succade': a form of early-ripening melon. 'Crasane': a pear said by Miller (1752) to be 'hollow'd at the Crown like an Apple: the Stalk is very long and crooked . . . the flesh . . . extremley tender, and buttery . . . the very best Pear of the Season [November]'—also known in White's time as the 'Bergamot Crasane' or 'Flat Butter-pear'. 'bason': holes filled with specially prepared soil, necessary at Wakes for many plants because of the clay soil. 'Battersea Cabbage-seed': Richard Bradley (1718: 149) mentions a 'large *White Cabbage*, of which the *Gardeners* at *Batersea* in *Surrey* have a Kind that comes very early . . . if we are minded to have of them ripe in the *Winter*, sow them in *March*'. 'borecole': kale. 'Holyoaks': *Althaea rosea*, hollyhock. 'Garden Lathyrus': everlasting-pea. 'p: sunflowers': perennial sunflowers, *Helianthus* sp. 'tree-prime-roses': *Oenothera biennis*, evening-primrose, known in White's time as night primrose. 'red Cowslip': a sport of *Primula veris*, commonly known as 'Devon red'. 'artichokes': not *Helianthus tuberosus*, Jerusalem artichoke, which is often listed in the eighteenth century with potatoes and was known earlier as 'Potatoes of Canada' (Parkinson, 1629), but *Cynara scolymus*, globe artichoke, the culture of which White refers to frequently (he also grew *C. cardunculus*, cardoon).

3. *Kalendar*, 13 Apr. 1759: James Knight's house is associated with Coneycroft, on the western edge of Selborne towards Faringdon; the ponds there, which attracted the swallows, were also the site where (in 1768) the common sandpiper *Tringa hypoleuca* was reported (see Pennant 20).

4. This work was first published in 1767 and ran to 25 editions, the last in 1848. The edition cited (now and later) is the third, and will be referred to as Abercrombie (1769). John Abercrombie (and not Thomas Mawe) was the author of the volume, although his name did not appear on the title-page until the seventh (1776) and subsequent editions. See Henrey (1975, **2**: 363–8).

5. The first recorded use for 'greenhouse' is in 1664, but by White's time it had become a generic term (see entries for 'Green-house' and 'Stoves' in Miller (1752)). Although White grew no exotics, he made extensive use of hotbeds and frames for starting tender plants, and improvised where he could. Evidently, he liked plants indoors: for instance, 8 Nov. 1758 is the note 'Set to blow in Glasses four Polyanth-Narcissus, & two Hyacinths', and earlier the same year, on 31 Jan., is the entry:

The Narcissus's planted in sand in common blowing-glasses, have crammed the glasses so full, that tho' they budded very strongly at first, they have hardly advanced at all since in height for many weeks: one of ye Glasses, that was crack'd by accident, is quite split to pieces by the large, strong roots. Took it out of the Glass, & planted it in a pint-mug fill'd with sand.

On several occasions fruit is ripened indoors: 'Cut-up a Cantaleupe that had been cut green, & laid in the Buffet [pantry] to ripen' (6 Nov. 1757); and use is also made of outbuildings. 'Potted-out two Pyramidal Campanulas [chimney bellflower], one with 14 stalks, & one with two' (28 June 1764); 'The Pyram: Camp: are drawn [up] by standing in the brewhouse: put them in the Alcove' (22 July 1764).

6. Abercrombie (1769) gives 24 herbs of this kind, that is, those used for flavouring food, and also refers to 'Physical herbs', which were grown for medicinal use. Bradley (1718: 160), in discussing 'Sallet-Herbs', those eaten as a salad, comments:

> I would have the Relish of the several Herbs consider'd, as for Example, which are most hot and biting; as *Cresses, Mustard, Sellery, Taragon, Chervil, &c.* and the more cool and insipid to the Taste, as *Turneps, Rape, Spinage, Corn-Sallet, Letuce, Purslane*, and such like; by this Knowledge a *Gardener*, when he composes a *Sallet*, will so judiciously mix his Herbs, that the too strong Taste of one Kind may not over-power all the rest, and discreetly make use of the insipid Sorts to qualifie the Pungency and Heat of the others, as the Season of the Year is more hot or cold; so that every *Sallet* we eat, may not only be agreeable to the Taste, but also a sort of Physick to the Body.

7. These figures are taken from a brief introduction (by John Clegg) to the 1975 facsimile edition of the *Kalendar*, but, since (especially with melons) White exchanged seeds and often named sowings by the year in which the seeds were saved, any figures should be taken as approximate.

8. A useful summary of White's attention to melons is Le Rougetel (1971: 412–18), although some of the data are imprecise: for example, the 'melon wall' (p. 415) was not for melons but for tree fruit, and was started and completed in 1761 (not 1759). White records: i. 'Began building my fruit-wall' (14 May 1761); ii. 'Finish'd my fruit-wall . . .' (25 July 1761). Further, although it is right to have drawn attention to 1759 (p. 414) as a good year for cucumbers, we know from the *Journal* that 1790 produced far greater quantities: see Note 10 below.

9. This was at Hartley House in the parish of Hartley Mauditt, due north of Selborne, White noting: 'Ananas are in cutting at Hartley' (20 July 1767). Sir Simeon was an MP for Hampshire 1761–79: his only reported speech was on a 'bill to prevent abuses in the packing or bagging of hops' — see biographical summary in Namier and Brook (1964).

10. An entry not in the *Kalendar* but from the *Journal* (31 Aug. 1789): it appears to be the highest figure for a single day. An addition of numbers in the period 16 Aug.–30 Sept. 1790 yields a total of 804 cucumbers.

11. This and the following three quotations are dated respectively 8 Sept. 1766; 17 June 1758; 1 Oct. 1765; 10 Apr. 1756.

12. Abercrombie's classification gives priority to whether the mature plant is best cultivated from seed or from vegetative propagation (by either cuttings or division), thereby emphasising the focus of his text on the growing of plants. Classification of plants as annuals, biennials, and perennials reflects, it may be argued, perspectives of ornament and display.

13. Abercrombie's 'First Class' annuals are fewer than a dozen, and include 'Balsamines', *Impatiens* sp.; 'Sensitive Plant', *Mimosa pudica*; 'Stramonium', *Datura stramonium* (thorn-apple). Of the 15 species of 'Second Class' annual listed, White grew most.

14. 'dwarf sunflowers': *Rudbeckia hirta* (black-eyed Susan); 'Marvel of Peru': *Mirabilis jalapa*, a plant whose flowers are best in the evening; 'China Aster': *Callistephus chinensis*; 'Fr: & Afr: marrigolds':

Tagetes patula and *T. erecta*, respectively; 'Convolvulus minor': *Convolvulus tricolor*.

15. Amaranths provide a simple but useful example of the caution one needs to bring to interpreting plant names. Before the mid-eighteenth century, the term included plants as various as prince's feathers, love-lies-bleeding, and cockscombs; but after Linnaeus (1753) the genus *Celosia* was formed (for the cockscombs), *Amaranthus* being retained for those in which 'the flowers of most of the species retain their bright colours when dead' (Loudon, 1829). Although Miller used this new classification for his *Dictionary* (1768), popular practice was, however, resistant, and both White and Abercrombie (1769) refer to cockscombs as amaranths. For a fine discussion of local, popular names of plants, see William T. Stearn, 'An Introduction to Vernacular Names', in Smith (1972: 339–51).

16. Part of entry under *Celosia*, with the addition: '. . . if the kind is good, (and there is no want of dung, or conveniences) in a kindly season, they will grow much larger'. But they are not to everyone's taste, and Christopher Lloyd (for example) notices the 'hideous foliage' and remarks that 'no artificial flower could look more synthetic' (*The Well-tempered Garden* (1985: 289).

17. China asters, *Callistephus chinensis*, were introduced in only 1731 (Loudon, 1829), Miller (1768) reporting that the plant is 'a native of China, from whence the seeds were sent to France by the missionaries . . . In the year 1731 [and in 1736], I received seeds . . . but these were all single. They came by the title of La Reine Marguerette, or the Queen of Daisies, by which title the French still call it'; he continues: 'In 1752, I received seeds of the double flowers both red and blue, and in 1753, the seeds of the double white sort . . .'. That White is growing the double form as early as 1755 (in 1754 he was busy still with the cockscombs) suggests that he may have obtained the seeds privately, rather than from a seedsman, but there is no note of this at the time of sowing, when he simply states: 'Made a slight hot-bed with 1 load of

dung for sunflowers, African-Marrigolds, double Asters, & Celeri; & hoop'd, & matted it' (23 Apr. 1755).

18. The difference in ground on the two sides of 'the cart-way', the main street in Selborne, is explained in Pennant 1: White's ground was 'a rank clay', the ground at the vicarage 'a warm, forward, crumbling mould, called *black malm*'. What variety of potato White grew is never declared, but, in contrast to the over 200 varieties available today (Lloyd, 1985: 430), there are only twelve given by Richard Townley, 'On the Culture of Potatoes', in A. Hunter (1777: 301–17).

19. See Simons (1942: 131).

20. *Selborne*, Barrington 37. For those of us brought up on the story of Raleigh's role in the introduction of the potato (and tobacco) this must seem a rather strange claim, but research bears out what White says. Although west of the Severn and north of the Trent potatoes flourished in the late seventeenth century, in southeast England it was not until mid-eighteenth century that much progress was made, potatoes until that time being associated with leprosy, venery and other ills. For a detailed and absorbing treatment of the potato, see Salaman (1949).

21. White must have been especially gratified at this success, since it appears to have been wholly attributable to his own example and encouragement, and even at his own expense (see *Selborne*, Barrington 37, where reference is made to the use of 'premiums', that is, to incentives). Other villages in southern England could have profited from White's guidance: at Horstead Keynes, where potatoes first arrived as late as 1765, no one knew how to deal with them, so a man from another county was invited over on each Lady Day, 25 March, to plant the villagers' tubers; but they were accepted only gradually, and 'at the elections which took place at Lewes about this period, [potatoes] shared with Popery the indignation of the people, and "No Popery, no potatoes!" was the popular cry' (Robert W. Blencowe, 'Extracts from the Journal and Account Book of the Rev.

Giles Moore', *Sussex Archaeological Collections*, 1848, **1**: 97).

22. Careful scrutiny of terminology used for sites suggests that it would be possible to reconstruct most of the garden, e.g. boundaries are often specified: 'broad borders under my Father's window', 'Border next ye street', 'Baker's Hill beyond the Field-Garden', 'borders in the yard', 'Lassam's hedge', 'next Parson's', 'new border under rod-hedge', 'house wall', 'orchard-walk', 'ditch by the sand-walk', 'towards Willis's', 'the Quick-set-hedge between Turner's, & the Orchard', and so on.

23. Exploration of data about village folk is best found in Emden (1956).

24. Most of the additions are made by White's brother Thomas, but the hands of his father and two other brothers (John and Henry) also appear.

25. More explicit information was included in the *Kalendar*. The most striking examples are two memoranda dated 12 Mar. 1752 which occur 15 pages later than the entry for the preceding day. Knowing he was shortly to leave for Oxford (to prepare for proctorial duty), White left instructions—for Thomas White—concerning the whereabouts of his 'new Cucumber-frames' (some parts were in 'the old barn', others in 'the lumber-garret', and 'the oaken-pins' were 'in a deal-box') and the quantity of manure needed in preparation of the beds: 'Seven *very full* cart-loads of dung make an exact suitable hot-bed . . .'. Not all such entries (marked as 'Mem:') were for other people: for instance, an entry comparing the survival in differing winter weather of cos lettuce (14 Feb. 1756) is clearly for White's own purposes; see also 23 Mar. 1767, and perhaps 17 Mar. 1755 although, knowing he was to be away from the end of April, this last may also be for those remaining at Selborne.

26. For some plants the crucial factor determining, in our English latitudes, a successful flowering and fruiting cycle is temperature (witness the pineapple: see Note 9 above); but for others, for example the chrysanthemum, the length of daylight or darkness is much more critical.

Although the importance of light to flowering behaviour was realised last century, the necessary research was established only around 1910 (by Julien Tournois), and generalised by Garner and Allard (1920). From this time on growers in America exploited the finding, but the technique was not adopted in Britain until the 1950s: for a useful review of research and practice, see Searle and Machin (1968). As for vegetables, it is economically much cheaper to freeze, can, and otherwise package (including flying-in) than to grow them fresh in local nurseries.

27. John [Beckhurst] is explicitly named as a gardener (5 Apr. 1757), see also Mulso (1907: 33, 36, 50). For Hoar, see Emden (1956). Help was also forthcoming from Berriman, 'Kelsey's people', [Robin] Tull, Farmer Knight, Will Dewey: 'Two labourers . . . hoed & weeded', 'Dame Turner, & Girls weeded all the brick walks'; at the time of moving his barn, '20 hands' assisted; and one summer 'four mowers cut [the grass on] the great mead, ye slip, & the shrubbery by dinner-time'.

28. As early as 1755 Selborne 'Yellow Cantaleupe-seeds' are saved (see 19 Mar. 1757); in the same year (25 Apr. 1757) there is a note about marking with sticks 'the finest blowers [of some polyanths] . . . to save seed from them'; similarly, on 4 Sept. 1761 White 'Mark'd the best, & most double annuals for seed', and earlier the same year (on 1 Apr.) 'a plot of stocks from seed of my own saving' was sown. Economy and pleasure (associated with aesthetic appeal and rarity) applied also to bulbs. Witness this entry for 30 Apr. 1763: 'Tyed those Hyacinths that are white with a pink-eye with a piece of scarlet worsted as a mark to save ofsets from'. There is also of course frequent mention of taking slips, and of layering.

29. The best examples here are: 'Sowed everlasting pease, & wild-Lathyrus [*Lathyrus latifolius*, broad-leaved everlasting-pea, and either *L. sylvestris*, narrow-leaved everlasting-pea, or *L. tuberosus*, tuberous pea] from the Lythe; soaked the seeds in water two nights, & a day' (16

Apr. 1759) — see also 31 Mar. 1761 for a similar occurrence; an entry for 21 June 1761: 'Discovered a curious Orchis in the hollow shady part of Newton-lane . . . It is the Orchis alba bifolia minor, Calcari oblongo [*Platanthera bifolia*, lesser butterfly-orchid] . . . I brought-away the flower, & mark'd the root, intending to transplant it into the Garden, when the leaves are wither'd'; and his account of the cranberry, *Vaccinium oxycoccus*, which he first saw 'on the bogs of Beans-pond in Wullmere forest . . . [but] could not venture on the moss to look after them', and yet returned four days later, for then there is the further note, 'Procured several Cranberry-plants from bean's pond with berries on them' (9, 13 Sept. 1765).

30. Identification of this item is intriguing. White is not referring to the seeds of a common, edible turnip, but possibly either to an ornamental arum with a tuberous root (probably *Arisaema* sp.) or to *Psoralea esculenta*; in this latter case, Sir Simeon Stuart would have obtained seeds prior to the date of introduction given by Loudon (1829) as 1811.

31. At Oxford, seeds were also obtained from the Physic Garden; and in London, Philip Miller was a source, e.g. 17 Feb. 1759:

> Received from Mr: Philip Miller of Chelsea about 80 mellon-seeds 1754: immediately from Armenia; which he finds to be better than those that have been first brought to Cantaleupe, & thence to England.

32. I have not been able to trace Middleton's nursery; John Williamson (*fl.* 1730–60) was a successor to Furber at Kensington Gore, and the nursery lasted well into the nineteenth century; from Armstrong and Forster at North Warnborough White first bought plants on 26 Oct. ('seven spruce-firs') in the year he began the *Kalendar*, 1751. For detailed accounts of the founders of nurseries, see Harvey (1974); Harvey (1972) should also be consulted. For a fascinating reprint of a later (1782) provincial garden catalogue, see Galpine (1983).

33. Which seedsman is not recorded, but the bulbs appear to have flourished since *The Calendar of Flora* (1766) gives details of their flowering: 'Crown-imperials . . . begin to peep [out of the ground]' (3 Mar.); 'Crown-Imperial . . . buds for bloom' (28 Mar.); 'Crown-imperial . . . blossoms' (14 Apr.); 'Crown-Imperials . . . out of bloom' (8 May); and later, in the *Journal* for 1791, they still blossom. One advantage of the plant is its height, Miller's *Dictionary* (1768) having the note:

> As this is one of the earliest tall flowers of the spring, it makes a fine appearance in the middle of large borders, at a season when such flowers are much wanted to decorate the pleasure-garden . . .

to which is added the comment that the plants are not seen as frequently as they might because 'the rank fox-like odour which they emit, is too strong for most people'.

34. In connection with this, consider the technical terminology deployed throughout the *Kalendar*: earthing up, staking, plashing, slipping, planting out, pricking out, stopping, trenching, basketing up, mulching, layering, thinning, rolling, grafting, and grubbing up are a few, all of which are still current.

35. For instance, he mentions sand, marl, Dorton mould, lime, soot, peat, coal-cinders, peat-ashes, tan, Blacksmith's cinders, earth cast up by moles, and so on: not all were used in the borders and fields (the tan was reserved for the hotbeds, and the mole-earth also was for particular purposes), but the variety indicates that the task of enriching and lightening the soil was tackled vigorously. For a brief description of the local soils, see 'The Geology of Selborne', an article by William Curtis included in Bell (1877, **2**: 374–7).

36. Evidently Hales had been working on this kind of proposal for some time since in this same year (1758) he published *A Treatise on Ventilators*.

37. Other examples of garden practice linked to homely resources include barrels sunk partly in the ground for water;

seeds stored in paper-bags; 'Crape-bags' tied to the best bunches of grapes as protection against wasps ('Mr: Snooke's grapes were eat naked to the stones a fortnight ago, when they were quite green'—8 Sept. 1762); crocus leaves saved as bines; birdlime and bottles of treacle and beer to catch wasps and hornets; and a home-made, free-standing, raised border—'Made a frame, or cradle for annuals of rods, & pease-haulm about four feet wide, & eight feet long; & put into it about 16 barrows of dung, & grass-mowings' (20 Apr. 1761). There were also boxes, pots, and mats; a sandstone roller; Dr Hill's 'mummy' (a rooting powder); and, of course, books. In this last field the main resource was Miller's *Dictionary*, but White also used Thomas Hitt, *A Treatise of Fruit Trees* (1755), which was followed carefully for pruning, especially his vines.

38. A craft blacksmith will still measure the temperature of his material by its colour, to indicate whether sufficient 'temper' has been reached for the tool to be used in a particular process; and the precise angle at which a skilled joiner holds his chisel to the sharpening stone is measured only by eye, 'feel', and (of course) practice.

39. The response of a liquid to gravity in combination with its own surface tension is a continuing source of measures. It has been pointed out to me (by Professor Burge) that laboratory benches are today levelled not by lengthy spirit devices, but by two water flasks joined by tubing, and that 'float' glass is so called because an even thickness is obtained by floating the molten glass on a bed of some other liquid material. White used a not dissimilar combination of gravity and surface ten-

sion to measure rainfall: 'very soft showers many times in a day; but not moisture enough to . . . make the eaves drop' (13 June 1765).

40. The stopping down of melons is at 4 May 1759; and the experiments with different waters for brewing begin at 9 Feb. 1763. Equally interesting are other occasions when he searches for an explanation to a gardening problem:

> Observing that the Succades were backward in setting, & went-off soon after blowing; I examined into the mould that lay on the lining, & found that it was so overheated by a thick coat of mowed Grass as to be scalding hot, & quite unfit for vegetation. Took-off the grass, & trod-down the earth close to the bed, where it was sunk away, watered it very stoutly, & fill'd it up to the frames with good fresh earth (26 May 1763).

The behavioural pattern that occurs here—an observation, an investigation, and then remedial action—is extremely close to characteristic modes White will employ frequently in connection with natural history writing: that we can see it at work first in the garden informs us (again) of his preparedness to become correspondent with Pennant, Barrington, and others.

41. In fact, the season sometimes militated against rainwater, which appears to have become a preference: writing to Thomas Barker as late as 1787, White remarks, '[I] always brew with rain water when I can' (Bell, 1877, **2**: 167).

42. Mention of a more significant example than any of these, as an indicator of White's concern for accuracy, occurs later (see Chap. 9, n. 44).

CHAPTER SIX

1. Over the years, this exclusivity has led commentators to write of White's insensitivity to national events, remarking (quite falsely in my view) on his evident isolation from matters of the greatest importance to the survival of a way of life

he espoused. Such opinions are a grave disservice to White. What needs to be explained is not the absence in *Kalendar* and *Journal* of quotidian familial, social and political occurrence, but their occasional presence. Hence, one can speculate what it was about the events that *are* recorded that prompted White to turn his eye momentarily from the task in hand (observation and record of the natural scene) to find himself so charged with interest and emotion that the steady flow of natural succession became interrupted.

2. Mulso (1907: 174).

3. Useful accounts of the life of the clergy at this time occur in William Addison, *The English Country Parson* (1947), and in *Johnson's England*, ed. A. S. Turberville (1952).

4. This and other details in what follows are given in White's accounts headed 'Expences from May 2, 1753. The day I went out of my Proctorship': see Bell (1877, **2**: 335–46). This set of accounts runs until 22 Mar. 1754.

5. One source for matters concerning White's college is Rannie (1900), but a fuller account of matters affecting White at Oxford is given in Scott (1946).

6. Mulso (1907: 94).

7. Ward (1958: 1), who explains something of the shifting alliances of the period.

8. Mulso (1907: 91).

9. In fact, so distressed that 15 years later he acknowledged a brother's good sense in not consulting a certain gentleman by remarking the man to have been 'a Whiting's man, & in those days a bigotted party man, & one in whose acquai[n]tance I had no complacency' (Foster, 1985a: 233).

10. Edmund Yalden's daughter had married White's brother Benjamin in 1753; and a son, William, was partner to White's brother Thomas.

11. The dates are given in White's accounts: see Note 4 above.

12. The only surviving register for this period at West Dene appears to be the Marriage Register, which shows that White officiated on 11 Aug., 14, 21 and 28 Sept., and 5 Oct. 1755 (County Record Office, Trowbridge, Wiltshire).

13. For the records themselves and a brief discussion, see Foster (1986a, **20**(1): 42–7).

14. *Selborne*, Pennant 30; Holt-White (1901, **1**: 85) cites the visits also. That

Barker was in Selborne at this time is known from his own diary, currently at Lancing College, West Sussex.

15. Mulso (1907: 101).

16. A useful measure of White's access to patronage is seen at the time his brother John sought a living in England to enable him to return from Gibraltar. White wrote 7 Sept. 1769 (the Miss Mulso mentioned being Mulso's sister):

> The husband's sister of Mrs Chapone (Miss Mulso) is married to a Mr Boyd a Gent: of large fortune in Kent: I am apt to suspect that this Mr: Boyd is yr Lieu: Governor's relation: & upon the strength of this Supposition have wrote a pressing letter to Mr Mulso desiring him to urge Mrs Chapone to en[t]reat Mrs: Boyd to desire Mr Boyd to write a letter to Governor Boyd in yr favour (Foster, 1985a: 230–1).

A circuitous route indeed, and one that proved abortive.

17. The memorandum-book of the Provost, for 15 Dec. 1757, reads: 'Moreton Pinkney given to Mr. White as Senior Petitioner, tho' without his intention of serving it, not choosing to waive his claim . . . I agreed to this to avoid any possibility of a misconception of partiality' (Scott, 1946: 81). The living itself was technically a perpetual curacy, there being neither rector nor vicar since the tithes were in lay hands.

18. See, for instance, Newton Valence Marriage Register for 2, 9, 16 and 20 Oct. 1757 (County Record Office, Winchester, Hampshire).

19. John White's will (Holt-White, 1901, **1**: 99).

20. Mulso (1907: 137).

21. The summary of White's income given by Mabey (1986: 75–6) suggests that his 'total annual income from sources other than his Fellowship (which earned on average about £100 a year) was at this time almost certainly little more than £100'. Fuller details are given in Holt-White (1901, **1**: 104–8).

22. Mulso (1907: 140, 143).

23. Details of this journey (and relevant

dates) are in an account book, headed 'Expences in Housekeeping at Selborn from the Death of my Father. Octobr: 1758', which runs to 28 Dec. 1762; this item is at the Gilbert White Museum, Selborne.

24. *Kalendar*, 17 May 1760.

25. There are numerous references to the idea of marriage in the Mulso correspondence. For example, when White was at West Dene, Mulso wrote: 'I sincerely wish You had a Living like Dene & the thoro' good Sort of Damoiselle that You mention, that your wishes might be compleated'; and eleven years later, in July 1766, he advised: 'These Summer Visitants of Your's are great Hindrances to a certain Scheme, which you once told me should be soon undertaken . . . I shall not let you go above a Year or two more, before I begin to take ye other Side of ye Question, & inveigh agst your undertaking this Yoke of Wedlock at all' (Mulso, 1907: 99–100, 203, respectively).

26. Over the years, Oriel offered White livings at Cholderton (Wiltshire) in 1764; Tortworth (Gloucestershire) in 1765; Cromhall (Gloucestershire) in 1767; Cholderton (again) in 1768; Cromhall (again) in 1774; Ufton Nervett (Berkshire) in 1779—which really attracted him, for he visited it twice to consider; and Tortworth (again) in 1785. See Mulso (1907: 193, 198, 206, 212, 248, 286, 329), and for the detail concerning Cromhall consult Bell (1877, **2**: 27); Holt-White (1901) is also useful.

27. Henley (who became Lord Chancellor in 1761) had been attorney-general but was elevated to Lord Keeper of the Great Seal in June 1757, and on 1–2 Oct. that year White noted in the *Kalendar*:

> Cut two very high-flavoured Cantaleupes, both under two pds in weight. They were very weighty for their size; & their coats very black, & emboss'd. Sent them to Lord Keeper.

The gift was repeated the following year (see 25 Aug. 1758).

28. The election for the chancellorship of the university was in 1759, and a full account is available in Ward (1958: 207–12). The government had chosen as their candidate a Tory, Richard Trevor (Bishop of Durham and sometime canon of Christ Church), but he lost the election to John Fane, Earl of Westmorland, a clandestine supporter of the Young Pretender and of the Jacobite interest generally: see summary in Sedgwick (1970). Although White himself was not a Jacobite, his sympathies were very much with the old order, and in spite of voting for Trevor it had been noted by Henley as being done 'in so ling'ring, cold, & disobliging manner, that he [Henley] could not but beleive [White] disinclined to any Services of that Nature': quoted by Mulso in a letter to White of 28 July 1763 (Mulso, 1907: 179).

29. Quotations from 'The Invitation to Selborne' given in Appendix D.

30. See Note 25 above.

31. Members of the family regularly visited Selborne (including in summer 1763 and winter 1763/4 White's aunt, Mrs Snooke, whose husband had died in 1763 only a few months after the death of Charles White, the vicar at Bradley); several of Mulso's relatives came also, especially in the mid-1760s; and in summer 1763 the Batty daughters were at the Vicarage, an event that led White to arrange Prospero-esque entertainments, the details of which are enjoyed by Mabey (1986: 88–94).

32. Mulso (1907: 196).

CHAPTER SEVEN

1. There is also a single page of entries dated 11 Aug. 1767. White gave the whole document the title 'Calendar of Flora', but, to avoid confusion with use of '*Kalendar*' for the *Garden-Kalendar*, this new item will be referred to as '*Flora*'. It was first published in facsimile, under the title *A Nature Calendar* (1911), by the Selborne Society, with an introduction by W. M. Webb.

2. The first edition of Stillingfleet's *Tracts* (which are translations of selected dissertations submitted by students of Linnaeus for degrees at the University of Uppsala) had been published in 1759; the 1762, second edition is an elaboration of the first, with the most important addition of 'The Calendar of Flora, Swedish and English' (that is, the Berger calendar and Stillingfleet's own), which was first published in 1761.

3. Stillingfleet (1762: 305).

4. For example, a bushel of wheat (enough, say, for thirty large loaves) cost 8s. 6d. (42.5p) at Windsor market on Lady Day (25 Mar.) 1757, but only 3s. 11½d. (19.8p) on the same day in 1755: see *Museum Rusticum et Commerciale* (1764, **2**: 130).

5. A useful survey of agricultural developments in the period is Fussell (1950).

6. Stillingfleet (1762: 234–5); preface to 'The Calendar of Flora'.

7. Stillingfleet (1762: 247–8).

8. Stillingfleet (1762: 255); Berger's introduction to his calendar.

9. *Selborne* (1789), Barrington 10 and Barrington 1 respectively.

10. Holt-White (1901: 31).

11. Partly this is because of the *Kalendar* entries, but also because in that year he began to use William Hudson, *Flora Anglica* (1762).

12. See, for example, Mabey (1986: 98), who claims merely that White began to botanise because he 'was in the mood for a new, absorbing, optimistic pastime'.

13. For example, as well as subscribing to a book-club at Alton (which occasionally exacted 'forfeits'), he bought Stephen Hales, 'statics' (*Statical Essays*, 1731–3); John Hill, 'herbal' (*The British Herbal*, 1756); Jan Swammerdam, *The Book of Nature* (1758).

14. To trace the development of White's intellectual progress in this whole area is extremely difficult in the absence of the kind of documentary materials that are available for, say, his contribution to the natural history of Gibraltar (see Foster, 1985a). What we do know (Mulso, 1907: 199–201) is that White had written about Stillingfleet early in August, and that Mulso responded: 'I have never seen the works of Stillingfleet that You speak of; & indeed it is a Subject that I am not so engaged in as Yourself . . . I have no Temptation to out-door work'.

15. Stillingfleet (1762: 39), from Isaac J. Biberg, 'The Oeconomy of Nature', but the phrase (or something like it) was frequently used in the century, and White would have come across it in one of the books he took to school at Basingstoke on 17 Jan. 1739, John Wilkins' *Of the Principles and Duties of Natural Religion* (1675, 7th edn 1722–3): see Holt-White (1901: 31–2).

16. The importance of this kind of 'coincidence' is noted by Stillingfleet (1762: 148): 'It is wonderfull to observe the conformity between vegetation, and the arrival of certain birds of passage.' See also Note 36 below.

17. I say 'about' rather than 'on' 9 Aug. deliberately: although White completed on this day one notebook of records about the garden, he also started another to which he gave the title 'A Calendar of Flora, & the Garden from August 9th: 1765'; but this is not the date when he first began to record wild as well as cultivated plants. For a full discussion of this textual issue, see Foster (1988).

18. It is easy to become bemused by modern demands for data that can be verified, or to think (for a past age) that the 'recording of localities for medically useful herbs' was of the highest priority; but interest in plant distribution and in the beauty and variety of plants 'for their own sake' is thought to have been pursued nearly a century and a half before White began his field studies: see Rose (1972).

19. For this quotation and those in the following sentence, see *Selborne* (1789), Barrington 40.

20. *Ranunculus flammula*, lesser spearwort.

21. For evidence that some of the early steps White took in botany were prompted by his brother Thomas, see discussion in an edition of the *Flora* currently in preparation by the present author in collaboration with Margaret Grainger.

22. White is very accurate in his recording here: an initial 'no down' was changed to the precise 'little down'.

23. For a detailed account of the introduction of Linnaean principles into English botany, see Stearn (1957).

24. Johnson (1928: 147–8).

25. *Selborne* (1789), Barrington 41.

26. Stillingfleet (1762: 145–8).

27. White's aesthetic response to plants is infrequent in the *Flora*, but there are other examples—for instance:

> 28 Apr: The uplands glow with pile-worts, & dandelions; the wet meadows with marsh marrigolds.
>
> 29 Apr: Laurustines in high bloom, & beauty still.
>
> 7 Aug: Broad-leaved spurge, tithymalus platiphyllos goes into pod. It is a tall, handsome plant.

For identifications of plants in the *Flora*, the glossary provided by Greenoak (1986) is useful, but see Chap. 8, n. 20.

28. This is a very selective list of what White notes: he records other stages for many plants, and there are several observations about the ripening of fruit, the maturity of catkins, the podding of leguminous plants, and the setting of seed.

29. Linnaeus's sexual system of plant classification needs, of course, plants in flower with stamens showing if they are to be ascertained with precision; but the focus of the whole notion of a natural calendar is illuminated well by Berger in Stillingfleet (1762: 257), who also explains the relevance of garden plants:

> Botanists and apothecaries, whose business it is to gather plants just when they are in blow, may by [keeping calendars] learn at what time that may be done, and need not seek in vain at an improper season, and may farther know by their garden plants, what wild ones are to be found in the fields precisely at the same time; and on the contrary.

30. In fact, White notes more than this: over 25 specific place-names are recorded, as are many habitats: for example, 'dry clover-fields', 'in wet ditches', 'on ye downs', 'on walls', 'sandy fields near the forest'.

31. Hence, White often records more than one local name: for example, on 7 May *Veronica serpyllifolia* L. is noted as 'Little smooth-speedwell, . or paul's betony'. Similarly with birds: 28 Oct. notes *Turdus iliacus* as 'The Red-wing, swine-pipe, or wind-thrush'. This is in fact a characteristic mode of naming, which is carried forward even into *Selborne*, where (in Barrington 2) White retains 'storm-cock' as a name used 'in *Hampshire* and *Sussex*' for the mistle thrush, *Turdus viscivorus*. As for insects, White was later (1770) to bemoan the absence of many common names, remarking in a letter to his brother John at Gibraltar (Foster, 1985a: 321) that 'few Naturalists understand insects well' and attributing the cause in part to there being 'no good English Hist: of Insects . . . & half the insects of this Island have no name in our language'.

32. See, for example, *Selborne*, Pennant 30:

> The *French*, I think, in general are strangely prolix in their natural history. What *Linnaeus* says with respect to insects holds good in every other branch: "*Verbositas praesentis saeculi, calamitas artis* [The verbal effusiveness of the present century is a disaster for art]."

33. 'The fruiting-Ananas are ranged in their beds to stand the winter: fires begin to be lighted in the stoves': this would have been at Sir Simeon Stuart's: see Chap. 5, n. 9.

34. Also, perhaps, to the time of their customary flight: in the evenings. For another crepuscular creature that fascinated White, the night jar, *Caprimulgus europaeus*, see Grainger and Williamson (1988).

35. Stillingfleet (1762: 236) comments:

> When men considered the wonderful

migration of birds, how they disappeared at once, and appeared again at stated times, and could give no guess where they went, it was almost natural to suppose, that they retired somewhere out of the sphere of this earth, and perhaps approached the aetherial regions, where they might converse with the gods, and thence be enabled to predict events.

36. An instance of this, echoing Stillingfleet, occurs in April. Stillingfleet himself (1762: 148) remarks:

> It is wonderfull to observe the conformity between vegetation, and the arrival of certain birds of passage. I will give one instance as marked down in a diary kept by me in Norfolk in the year 1755. April the 16th *young figs* appear, the 17th of the same month the *cuckow* sings.

For White in 1766 the association was:

> 15 Apr: The vine, vitis, budds.
> Polyanth-Narcissus blows.
> Young figs appear.
> 16 Apr: [No entry]
> 17 Apr: Nightingale, luscinia, returns, & sings.

This may not seem remarkable, but White was clearly taking the bird as especially prognostic and the entry is made in a script almost twice the size of the surrounding matter. Belief in the prognostic power of birds is still current in the present century: at the time of the severe summer of 1976, John Crompton (a farmhand at Llanynis, Builth Wells, midWales) remarked to the present writer that he had observed that when the swallows arrived in May they began building nests immediately, whereas normally they delayed such activity for about a fortnight. On my asking why this might be, Mr Crompton said that if the birds had followed custom there would have been no nests at all because all the available mud would have dried out 'and they surely knew that'.

37. One of the most appealing is the glow-worm, *Lampyris noctiluca*, which White may have taken as prognostic. At the end of the 14 June entry (after eleven other items) occur two further items, both written in larger script than that on the rest of the page:

> The glow-worm appears.
> Wheat ears begin to peep.

As for insects, there are several species of bee, butterfly and fly, on some of which White writes in detail. For example, part of the entry for 17 Oct. reads:

> Musca apiformis, tota fusca, caudâ obtusâ, ex culâ caudatâ in latrinis degente orta, still is seen. This fly frequents sinks, & jakes, where it lays it's eggs.
> In the autumn it feeds on the flowers of late annuals, & perennials; & in particular on the blossoms of Ivy. Ray hist: Insect: 272.
> Musca bipennis major, diversicolor, caudâ setis nigris obsitâ appears, & engenders. This fly is entirely a garden or field-fly, never entring into houses: it appears to feed on mellow fruit. Mr: Ray seems not to have been aware that it smells strongly of musk: it might therefore·not improperly be call'd musca moschata. 271. This seems to be an autumn fly altogether.

38. One way in which this is shown is the pleasure White takes in what is new to him, and this will be demonstrated especially in the response he makes to the specimens that will shortly inundate him from Gibraltar. As for man's ability to unlock the secrets of the natural calendar, hints of scepticism begin to enter the record. For example, at 24 Oct. there is a long list of 'Plants naturally in bloom still' followed by one almost as long of 'Plants continued in bloom by accidents, such as a shady situation, the bite of Cattle, &c:'. Further, at several points there are notes that record two stages of growth: for example, at 20 June 'Yellow cow-wheat [*Melampyrum pratense*] . . . in flower, & pod', and on 8 Sept. 'Small burnet-saxifrage [*Pimpinella saxifraga*] . . . blows, & seeds'. Other notes include 13 Mar: 'This [Pilewort] & other things are forced

before their time by the hot sun shine'; and a note at 4 Apr. about the early awakening of Mrs Snooke's tortoise at Ringmer. Of course, plants that proved unreliably prognostic could be excised from the record, and ignored; but the observations White made over the years, and the way he made his records, never quite suggest that that is the conclusion he should reach.

CHAPTER EIGHT

1. I refer to White's records alone here; on a broader scale, James Jurin (Secretary of the Royal Society) had invited observers to submit standardised reports to the Society as early as 1723. The most important English records for the century are those of White's brother-in-law, Thomas Barker: see Kington and Kington (1981: 58-67).

2. A popular, contemporary compendium of folk guidance is a volume known as *The Shepherd of Banbury's Rules* by John Claridge: it was first published in 1670 (under the title *The Shepherd's Legacy*), and was available throughout the following century, especially in an elaborated edition of 1740. A more recent edition of 1946 with a useful introduction by G. H. T. Kimble is a reprint of an 1827 edition.

3. Records in the *Kalendar* and 'An Account of the brewings of strong-beer. An Chronicle of strong-beer, & raisin-wine' (at Selborne Museum) show that White prepared annually about 20 gallons of wine, made from raisins and enriched usually with elderberry syrup; he also brewed strong beer, often about 50 gallons over a year, in spring and early winter when the temperature was equable enough for the fermentation, and sometimes ale and small beer as well.

4. The more obvious places where White shows his awareness of this are in *Selborne*: Pennant 18 reveals his anger over a quack cure for cancer by means of toads, and Barrington 28 discusses the power of superstition over those whose 'minds are not invigorated by a liberal education'.

5. The contribution of many members of White's family both in his own and in the other immediate generations will repay further study. Just as the family proved to be so notably loyal to Oriel College (13 Whites were admitted between 1704 and 1870: see Emden, 1948: 218), so too were they loyal to natural history. Of all the contributions, that of Thomas was probably the most significant; witness this item in his Commonplace Book (now at the Selborne Museum):

> Butomus umbellatus. This plant grows in watr'y Ditches & sides of Rivers, & would be very ornamental if it was introduced into any suitable place, where the owner is willing to add to the natural productions, but at present we do not either, cultivate or ornament watry places if they are situated too low for draining; the Chinese have carried their improvements so far, that instead of draining Land, they introduce such plants as are fit for food, & thrive in the Water.

6. British Library Add. MS 35138, White to Pennant, 10 Aug. 1767.

7. A contemporary encomium reads:

> Originally partner with Mr. John Whiston ... Mr. White afterwards opened a separate shop ... particularly in the line of Natural History, and other expensive books. He retired from business with a plentiful fortune ... Benjamin, his eldest son, retired also in a few years after him; and is still living at Hampstead; leaving the business to a younger brother, John, who is also about to retire, with an easy competence, to the enjoyment of a country life (Nichols, 1812: 127).

The success of the business is, of course, partly revealing of the popularity of natural history as a field of interest in the period.

8. See, for example, Noblett (1982: 61–8).

9. B.L. Add. MS 35138, White to Pennant, 10 Aug. 1767. The terms of this extract seem clear enough: that White and Pennant met personally, albeit briefly and casually, and exchanged remarks about natural history. Holt-White (1901, **1**: 152) questions whether this was so, adducing evidence from a much later letter (28 Feb. 1769) in which White hoped that, since 'we [White and Pennant] have been so well acquainted by a long and communicative correspondence, I trust we should relish each others conversation, and be soon as well acquainted in person as by letter'. However, to sustain the view that the correspondents had not met in 1767 leads to a complex series of hypothetical encounters: for example, that White's observations (which? in what form?) were communicated by White's brother Benjamin to Pennant, although they were unsolicited; that Pennant acknowledged their merit to Benjamin; that Benjamin, although noted as being a poor correspondent to the family, passed this recognition on to White; and that White, on that basis alone, was willing to contemplate writing, and actually did write, to a perfect stranger about his own interests. All this seems convoluted—and unlikely.

10. White's description of this mouse is the earliest known, and its discovery and report (to Pennant, who, acknowledging White, included a long account in his 1768 edition of *British Zoology*: see vol. 2, pp. 498–500) are one of his most important contributions. His own description was not published until 1789 (in *Selborne*) and he is not therefore credited with priority as regards the creature's name: this belongs to P. S. Pallas, who found the mouse on the banks of the Volga and published a description in *Reise durch verschiedene Provinzen des Russischen Reichs*, 3 vols. (1771–6).

11. Analysis of White's known reading is not yet available, but a list of the main authorities in *Selborne* is compelling evidence of familiarity with most of the major texts in natural history of his age: for example, as well as Ray and Linnaeus, he refers to Adanson, Aikin, Banks, Belon, Brisson, Buffon, Derham, Edwards, Ellis, Fothergill, Gassendi, Geoffroy, Hales, Hasselquist, Huxham, Kalm, Lightfoot, Lister, Miller, Plot, Reaumur, Stillingfleet, Swammerdam and Willughby. And this is not all his reading: he read widely in poetic literature (as seen earlier), was fluent in some of the main texts of the classical writers, and had a penchant for works of history.

12. Interestingly, most of these remarks occur in the later series of letters in *Selborne* addressed to Daines Barrington: for example, Barrington 3 ('Many times have I had the curiosity to open the stomachs of *woodcocks* and *snipes* . . . [to discover] what their subsistence might be'); Barrington 5 (where he proposes that 'by opening a female during the laying-time' it may be determined whether the cuckoo 'lays one or two eggs, or more, in a season'); and Barrington 31 (which reports the cutting-up of a female viper, and the discovery that 'the abdomen was crowded with young, fifteen in number'). In the Pennant letters the most famous measurements are those in connection with the moose at Goodwood: see Note 23 below.

13. B.L. Add. MS 35138, White to Pennant, 22 Jan. 1768.

14. Yet on occasion some historical sensitivity is required: in the quotation just given the mention of 'winter quarters' refers to the contemporary practice of soldiers being billeted in villages and parishes for the winter, and then assembling in the spring (when the weather and transport conditions permitted) for a campaign. Report of such an event appears in a letter of 22 Jan. 1783. Writing to a relative, White remarked:

We have had all this winter 26 Highlanders of the 77 regiment quarter'd in this village . . . where tho' they had nothing in the world to do, they have behaved in a very quiet, & inoffensive manner; & were never known to steal even a turnip, or a cabbage, tho' they lived much on vegetables, & were astonished at the dearness of Southern provisions. Late last night came an express ordering these poor fellows down to Portsmouth; where they are to embark for India (Rylands MS 1306/23).

The soldiers' arrival at Selborne was unplanned. Apparently, in the previous autumn, they

> were embarked in the S: of Ireland in order to have attended Ld Howe to Gibraltar: but a cross wind drove them to the back side of Cornwal, & so to Ilfracomb on the N: of Devon, where they were landed (White to Molly, daughter of Thomas White, 14 Dec. 1782, Fitzwilliam Museum, Cambridge).

15. These included: 'Marsh St: Peter's wort with hoary leaves . . . Purple marsh-Cinquefoil . . . Water dropwort . . . Marsh-goose-grass'.

16. B.L. Add. MS 35138, White to Pennant, 9 Sept. 1767. Several lists of plants growing in the area of east Hampshire have been prepared over the years: one of the most interesting is a pamphlet produced by the Curtis Museum at Alton (Childs, 1949).

17. B.L. Add. MS 35138, White to Pennant, 6 Nov. 1767.

18. B.L. Add. MS 35138, White to Pennant, 22 Jan. 1768.

19. Although the *Journal* went through several issues, the 1767 issue (which carries the imprint 'LONDON: Printed for w. SANDBY, in FLEET-STREET') is discreetly anonymous as to authorship. Later, Benjamin White published it and Barrington permitted his name to be given at the end of the 'Preface'. White's accounts (at Selborne Museum) show that in the 1780s he bought his copy from Benjamin annually and paid five shillings for copies halfbound and interleaved.

20. To understand Barrington's purpose in devising the *Journal* it is vital that this tripartite structure of the layout of each page be grasped. Sadly, this is not apparent in Greenoak (1986), where the General Editor (Richard Mabey) has approved a design which reduces the data central to the whole endeavour to the status of marginalia and offers as a main text notes in the column headed 'Miscellaneous . . .'.

21. Quotations here and earlier from the *Journal* are taken from the Preface as well as from headings of columns (the printed text was not perfect in this first issue: for example, some pages have columns transposed).

22. This and the previous quotation are from B.L. Add. MS 35138, White to Pennant, 22 Jan. 1768.

23. Several moose were imported to Britain in the late 1760s and created considerable interest. White heard of the Goodwood specimens through Pennant and responded:

> Your account of the Moose gives me a great deal of satisfaction; not only because I am glad to hear that two such animals, so little known, are arrived in this neighbourhood: but because in it You give me Hopes that I may have the Honour of yr Company at Selborne: & I earnestly desire that You will not disappoint me of that satisfaction. Tho' the direct way to Goodwood from Town is down the Chichester-road: yet if you will come the Alton [road], & so to Petersfield, there will be but a very few miles difference: & in yr way to Petersfield you will pass within three miles of my House: & my Horses shall meet you on the turnpike to carry You to this place (B.L. Add. MS 35138, 14 Mar. 1768).

At this time White had already been told of the creature's existence by a friend, James Gibson, who had seen service at Quebec (1759) as a naval chaplain (Foster, 1986c); this was the same friend from whom he often obtained garden plants at Bishop's Waltham. For discussion of contemporary interest regarding the moose, see Rolfe (1983).

24. B.L. Add. MS 35138, White to Pennant, 19 Apr. 1768.

25. Scott (1946) provides useful insights into Oxford concerns, but, for a biography focused much more on the

familial, Lockley (1954) should be consulted.

26. B.L. Add. MS 35138, White to Pennant, 10 Aug. 1767.

27. On 5 May 1767, Banks wrote to Pennant: 'it does not go on with the Spirit it used to do when you was with us' (Alexander Turnbull Library, Auckland, New Zealand, ALS 269). At this date Banks was just 24 years of age: his family held large estates at Revesby, Lincolnshire, and he had been educated at Harrow, Eton, and Christ Church, Oxford (where he hired a tutor to instruct him in natural history); he had been elected FRS in 1766.

28. Alexander Turnbull Library, ALS 269; but see Beaglehole (1977), who prints part of a further letter from Banks to Pennant of 14 May 1767 which refers to White in favourable terms.

29. B.L. Add. MS 35138, White to Pennant, 14 Mar. 1768.

30. Bell (1877, **2**: 241–2).

31. B.L. Add. MS 35138, White to Pennant, 16 June 1768. Solander (1733–82), Assistant Keeper at the British Museum, was a distinguished pupil of Linnaeus, and was to accompany Banks on the 'immense voyage'. White was in London a bare ten days, leaving Selborne on 3 May and returning on 14 May (see *Journal*, which for this period notes the availability of herrings, mackerel, sturgeon, and 'Green gooseberries').

32. Cook's *Endeavour* voyage was primarily a scientific and political venture: to observe the Transit of Venus, and to search for a great southern continent which, if claimed for Britain, might affect favourably the balance of power in Europe. To the expedition, Joseph Banks, a young and wealthy Fellow of the Royal Society, was permitted to add—at his own expense—a party of nine, who would record botanical and zoological discoveries. The expedition sailed from Plymouth on 25 Aug. 1768, Banks then being 25 years old.

33. B.L. Add. MS 35138, White to Pennant, 8 Oct. 1768.

34. B.L. Add. MS 35138, White to Pennant, 25 Sept. 1771. The following year White received a graphic account from William Sheffield about the materials assembled by Banks during the voyage (Bell, 1877, **2**: 97–100). For an elegant account of many aspects of the voyage, see Carr (1983).

34. B.L. Add. MS 35138, White to Pennant, 16 June 1768.

36. The natural history portion has the letters to Pennant and those to Barrington, but because they are directed to an unnamed recipient (although there was some discussion that they might be directed to Dr Chandler: see Chap. 12, n. 9) it is often forgotten that the letters of the 'Antiquities' are also a set.

37. Chardin Musgrave, White's vanquisher as provost in 1757, died on 29 Jan. 1768 and John Clark was elected in his stead on 12 Feb. 1768, White (as Senior Fellow) travelling from London (see *Journal*) to participate; he was not a candidate himself on this occasion.

38. George Huddesford was a long-standing member of the university, having been elected President of Trinity in 1731; he was Vice-Chancellor at the time of the famed county elections of 1754 (Emden, 1948: 102–3). His son, William (b. 1732), was a noted antiquary and at the time of an early death (1772) was Keeper of the Ashmolean.

39. Mulso (1907: 215–16).

40. B.L. Add. MS 31852, White to Barrington, 6 July 1769. The letter as printed in *Selborne* is substantially the same except for the exact detail of dates, since White then changed the time he was in London to 'last month' and the date of his letter to 30 June. The reasons for changes of this kind are reportedly obscure, but as here can usually be elucidated. When revising his letters for publication, White quite regularly, and, for the period, properly, deleted or softened much detail of a personal kind: here, faced with changing the month of May to something less precise, and choosing 'last month', a modification to the date of writing is automatic, and White elected for the date nearest to the actual date of the original letter. More generally, it is worth noting that dates of observations in

connection with natural history are in the main unchanged (but see Chap. 12).

41. B.L. Add. MS 35138, White to Pennant, 1 Sept. 1769, in the hand of an amanuensis with White autograph additions. That this is an original letter (sometimes White had his correspondence transcribed to keep a copy) is shown on the last page, which carries a Pennant autograph memorandum indicating that he answered the letter on 1 Oct. 1769.

42. Letters from Skinner to White are currently being prepared for publication by the present writer; for letters from Sheffield to White, see Foster (1985b).

43. This report first occurs in White's fourth letter to Pennant of 22 Jan. 1768 (B.L. Add. MS 35138), which is a response to Pennant's suggestion that many summer migrants to Britain overwintered in Spain (White himself having raised the whole question of migration in his third letter, of 6 Nov. 1767). White wrote on 22 Jan:

> What you suggest with regard to Spain is highly probable. The winters of Andalusia are so mild, that in all probability the soft-billed birds may find Insects sufficient for their support. I have a relation, an intelligent person, at Gibraltar, who often takes an airing into Spain: but he has never turned his thoughts to these studies. Tho' I remember in one of his letters a copy of verses in which it was implyed that the nightingale came to their rock at a stated season in the spring, & then left them.

As a result of this news Pennant pressed White to enlist his brother's help, and later in the year (letter of 28 Nov. 1768) White responded to Pennant by saying that he had written to his

> South country correspondent at Gibraltar, & urged him to take up the study of Nature a little; & to habituate his mind to attend to the migrations of birds & fishes; & to the plants, fossils, & insects of that part of the world.

44. A text of White's 14 letters to Gibraltar, from the first (Sept. 1769) to the last (Mar. 1772), is available in Foster (1985a).

45. These are White's only known correspondents at this time: there may well have been others who no longer appear in the record, and later he communicated with other naturalists, for example, Thomas Griffith and John Lightfoot, both of whom visited Selborne in the early 1770s.

CHAPTER NINE

1. There are several ways of counting the number of entries, the most difficult problem being what to specify as an 'entry'; and White's practice varied over time, especially when he began to interleave the *Journal*. Many species are entered twice the same day (once in Latin, once in English), and wind data are indicated when there is no wind by the absence of a symbol (which also records the omission of observation); so the position is complex.

2. In fact this is what is being done: see Kington and Kington (1981).

3. The standard text is still Townsend (1904), and an interesting contribution in the field is Vaughan (1906).

4. See, for example, A. Hunter, *Georgical Essays* (1777), which includes essays reporting experiments on times of sowing, and on the preparation of soil. A valuable guide to other relevant material is *Catalogue of the Walter Frank Perkins Agricultural Library* (Southampton University Library, 1961).

5. Data for 22 Sept. 1770 are a good example: there are two readings for temperature and two for pressure, but from the placement and from White's provision of location we know that the first of each double reading was made at Fyfield (about 36 miles west of Selborne, near Andover, where brother Henry lived) and the second at Selborne.

6. Apart from Selborne, White knew best the country at Fyfield and at Ringmer, both of which places he visited once or twice a year; it was travelling to the latter that led to some knowledge of coastal scenery and habitats.

7. The proposal went via the son of brother Thomas. Writing to the latter on 29 Nov. 1783, White remarked: 'If Mr. Nichols [editor of *The Gentleman's Magazine*] approves of my *diaries*, he may have more; they may serve to oppose to y^r father's, kept near town' (Bell, 1877, **2**: 144). Work on this aspect of White studies has only recently been started: see Sherbo (1985).

8. Foster (1986a: 320, 327). As for White's readings, they were regularly taken from an instrument kept in the grounds of Wakes: the *Journal* entry at 3 Nov. 1782 has in the column for rain '145', and, in an adjacent note, 'Rain on the tower 138'. It should also be noted that columnar rain records appear always to relate to conditions at Selborne, White usually adding when absent from his home a note such as 'Thomas [Hoar, his servant] kept the rain account at Selborne' (*Journal*, June 1780).

9. John Huxham, M.D., *Observationes de Aëra, et Morbis epidemicis, ab anno 1728 ad finem anni 1737 Plymuthi factae* (London, 1739); close scrutiny of Huxham by White is referred to in a letter of Dec. 1782 (Holt-White, 1901, **2**: 85–6).

10. Bell (1877, **2**: 167, 173, 176) gives details of some of these summaries; and there are others.

11. That White's thermometer should be marked in the Fahrenheit scale is not as obvious as one might think. Details of the scale had been published by the Royal Society in 1724, but there were many other scales in use throughout the century. The need for an agreed scale was urgent, as appears from a thermometer at Utrecht dated 1754 which is mounted on a board showing 18 different scales, none of which is centesimal (Knowles Middleton, 1966).

12. In the first edition of *Selborne*, which gives only 65 letters to Barrington, two consecutive letters are numbered as Letter 61. In fact there are 66 letters, and my references, here and elsewhere, use corrected numbering.

13. This was in 1803 in an essay by Luke Howard, 'On the Modification of Clouds . . .' (*Philosophical Magazine*, XVI). An earlier description of cloud formations can be found in the history of painting, and I am indebted to June Kessell, who has pointed out the work of Alexander Couzens (1717–86), '*Method to Facilitate the Invention of Landskip*, (1759), in which he included numbered illustrations of landscapes under different weather conditions . . . [and in] 1785 . . . compiled a schematic cloud-book' (*Shock of Recognition*, a catalogue of 'The Landscape of English Romanticism and the Dutch seventeenth-century school', The Arts Council of Great Britain, 1971).

14. This is so for a single year, 1770; an analysis of whether other years yield the same answer is awaited.

15. B.L. Add. MS 35138, White to Pennant, 25 July 1768.

16. B.L. Add. MS 35138, White to Pennant, 28 Nov. 1768.

17. In fact, for most of these creatures there are only the occasional comments such as those on cats: 'Cats catch all the birds that come-in for shelter from the cold' (18 Jan. 1776); 'Cats often catch Swifts as they stoop to go up under the eaves of low houses' (6 May 1780); 'The cat gets upon the roof of the house, & catches young bats as they come forth from behind the sheet of lead at the bottom of the chimney' (15 July 1786).

18. Most of these were needed to sustain human life; a detailed account of one of them, the extent of rabbits in maintained warrens, is Tittensor and Tittensor (1986).

19. *Selborne*, Barrington 11, dated 8 Feb. 1772; see also Barrington 14.

20. Ray (1750: 110).

21. Miller (1768) thought, for example, that orchids, 'for the extreme oddness and beauty of their flowers, deserve a place in every good garden', and that they could be transplanted easily if their roots were marked and they were moved into appropriate conditions when the leaves

were decayed—which is a practice White followed (see Chap. 5, n. 29). What success he had is difficult to judge, but it was markedly less than that of a later naturalist whose 'grand total [was] at least sixteen' (Grainger, 1983: 295–7).

22. Foster (1985a: 491).

23. The *Journal* for 24 Nov. 1770 has the note:

> The wild wood-pigeon, or stock-dove begins to appear. They leave us all to a bird in the spring, & do not breed [here] . . . The numbers that come to these parts are strangely diminished within these twenty years. For about that distance of time such multitudes used to be observed, as they went to & from roost, that they filled the air for a mile together: but now seldom more than 40 or 50 are to be anywhere seen.

24. The first authoritative judgement on this appears to have been over seventy years after White's first observations:

> The Ring Ousel is a summer visitor to the British Islands . . . [and] arrives in this country from the south . . . White of Selborne saw them frequently when on their route in Hampshire and Sussex . . . flights probably go to France and Spain, and from thence to North Africa, where they pass the winter (Yarrell, 1843, **1**: 207, 210).

25. *Selborne*, Pennant 34, part of a letter from Mar. 1771.

26. Notes on the species separation occur in the *Journal* on several occasions (for example, 6 Jan and 22 Dec. 1769; 25 Jan. 1770), and comment is included in an early letter to Pennant, that of 22 Jan. 1768 (Pennant 13), but the main speculation is to Barrington in a letter of 20 Dec. 1770 (Barrington 8):

> There are doubtless many home internal migrations within this kingdom that want to be better understood: Witness those vast flocks of hen chaffinches that appear with us in the winter without hardly any cocks among them. Now was there a due proportion of each sex it should seem very improbable that any one district should produce such numbers of these little birds; & much more when only one half of the species appears: Therefore we may conclude that the Fringillae coelebes [chaffinch] . . . have a peculiar migration of their own in which the sexes part (Selborne MS).

Male chaffinches, we now know, leave the flocks in late winter to seek breeding territories.

27. For 'blue mist' see Appendix E. White maintained a keen interest in meteorology throughout his life: as he wrote to his nephew Samuel (son of Thomas Barker) in 1786, 'I love to study climates' (Bell, 1877, **2**: 161). With this same nephew he exchanged accounts of weather phenomena (see particularly Bell, 1877, **2**: 126, 140, where he provides Samuel with data obtained from J. R. Forster, naturalist on Cook's second voyage, 1772–5, about southern auroras), and also sightings of the planets: 'Venus, Jupiter, Mars, and Saturn appear now every clear night, as it were in a line', he wrote on 30 Mar. 1775, and he twice reported seeing Mercury, which is particularly worthy of record since Kepler is reputed to have remarked that he never saw that planet in his whole life.

28. Before 1776, occasional notes can also be seen across column boundaries, and the margins of the paper, especially in a footnote position, were used from the earliest days. Additional support for 1776 being an important year for White's practice is found in interleaving: the first blank sheet is tipped-in between 25 and 26 May that year. The first fully interleaved *Journal* (a 1767 Sandby first edition; although in the late 1770s White used the 1775 Benjamin White second edition) is that for 1781.

29. Apart from sustained comment on the weather or some gardening problem, the most notable comment is the long (about 600 words) account of 20 May 1761 about the field cricket, *Gryllus campestris*, which White and his brother Thomas studied in the Short Lythe at

Selborne, and which White hoped—after moving some to his garden—would increase 'on account of their pleasing summer sound': and see *Selborne* (Barrington 46), where he evidently took one indoors and reported that

> when confined in a paper cage and set in the sun, and supplied with plants moistened with water, [it] will feed and thrive, and become so merry and loud as to be irksome in the same room where a person is sitting.

For White's letters on the house cricket and the mole cricket, see Barrington 47 and 48.

30. The ones not instanced in the text concern the singing of the robin; a query about the use of the term 'fructification'; comment on the similarity of young cock and hen chaffinches; suggestions about how best to gain further data on a new form of oak, and to verify whether swifts return year after year to the same nest; confirmation that some female avians posture on receipt of food from males; explanations of why shell-less snails seem not to hibernate, and why birds eat red currants before white; and a request for verification regarding a caged cuckoo never uttering 'cuckoo'. Commentary on each of Barrington's annotations can be found in Foster (1986b).

31. This was reported by White to Samuel Barker in a letter from London of 7 Feb. 1776: 'Mr. Barr. wants me to join with him in a Nat. Hist. publication' (Bell, 1877, **2**: 119). It seems probable that Barrington made the suggestion after savouring White's 1775 *Journal*, which White brought to London in late January 1776.

32. He does, though, write about migration (from a position of scepticism) and, occasionally, about trees.

33. White includes comment on which birds wash and which take a dust bath in Barrington 7 (8 Oct. 1770), suggesting that purification rites in Islam might derive from observation of this avian practice. It is likely that he came to such matters from travellers' tales, but his interest might have been encouraged by brother John's experiences in southern Spain and North Africa. Very much later (1786) there is evidence that he knew something of the Koran (which had been first translated into English in 1734) from first-hand readings: 'I believe all fervid regions afford instances of undulating vapours [he wrote to a nephew], that at a distance appear like water [mirages]: Arabia I know does; and the phenomenon is finely alluded to in the Koran' (Bell, 1877, **2**: 161).

34. Summer dispersal is recorded on several occasions, and White found his 1776 observations especially intriguing: 'Some of the little frogs from the ponds stroll quite up the hill: they seem to spread in all directions', he recorded on 10 July, and followed this on 14 July with: 'Young frogs migrate, & spread around the ponds for more than a furlong: they march about all day long, separating in pursuit of food: & get to the top of the hill, & into the N: field.' Earlier, in 1771, he recorded emergence and dispersal on 20 Apr., which, for a cold season (as 1770 and spring 1771 were), seems rather unlikely; in this instance his froglets may have been young smooth newts, which commonly emerge in spring.

35. Bell (1877, **2**: 25, 62).

36. And this extended to accounts of his journey to North Africa: 'Y.̣ Embassy to Morocco . . . will make a good chapter in your [natural] history' (Foster, 1985a: 235).

37. The years 1770 and 1774 were particularly bad for the common cockchafer or maybug, *Melolontha melolontha*. In the former year the *Journal* notes for 30 June: 'The Rooks pursue & catch the chafers as they flie. Whole woods of oaks are stripped bare by the chafers'; and in the later year, for 1 July, is the reflection: 'When Oaks are quite stripped of their leaves by chafers, they are cloathed again soon after midsummer with a beautiful foliage: but beeches, horse-chest-nuts, & maples, once defaced by those insects, never recover their beauty again for the whole season.'

38. In this sense, studies of the physical world (including, for example,

astrophysics) share the indeterminacy of many biological processes. 'Weather systems and cosmic nebulae are . . . not in principle different from living systems. The number of potentially possible interactions [stochastic perturbations] in such highly complex systems is far too large to permit prediction as to which one will actually take place' (Mayr, 1982: 42).

39. There are numerous factors that affect the precision of instrumental readings, but for White's data the most important would seem to be the location of his thermometer and the quality of the mercury used. As regards location, we should note that in his first *Journal* of 1768 he added to the column for temperature the word 'Within' and also the maker's name 'Ben: Martin', and at occasional points later numerical data are accompanied by the word 'Abroad', namely 'outdoors'. It is safe, then, to say that unless otherwise guided by White all his data for temperature related to an indoor location. However, at least two sites are named: in the tabulated data for Dec. 1782 we are told 'The thermometer hangs within an open staircase' (Bell, 1877, **2**: 143), and in *Selborne* (Barrington 61), where he reports the severe frost of Jan. 1768, he affirmed that the readings were taken from a 'thermometer within doors, in a close parlour where there was no fire'. Any detailed analysis of White's records would need to consider whether some allowance for this kind of variance should be made, especially since readings taken indoors in winter (at certain times of day) would probably be higher than the temperature pertaining outdoors, whereas in summer the opposite effect might be shown.

With the barometer, White knew something was likely to be wrong. The *Journal* has at 11 Feb. 1773: 'Reduced my barometer to the true standard of 28 inches, lowering it about two degrees'; and in 1790, early in October, is the comment:

It is much to be wished that all persons who attend to barometers would take care to use none but pure distilled Mercury in their tubes: because Mercury adulterated with lead, as it often is, loses much of it's true gravity, & must often stand in tubes above it's proper pitch on account of the diminution of it's specific weight by lead, which is lighter than mercury.

40. White cultivated *C.maritima*, sea-kale, for several years, having brought plants back with him to Selborne from Devon in 1750. As for *A. donax* (a bamboo-like grass of southern Europe which can attain 5 metres), White asked his brother at Gibraltar for seed in 1771 (Foster, 1985a: 490), noted in the *Journal* at 30 Sept. 1775 that it had grown that year 'eight or nine feet high . . . [, and] . . . opened the head of one stalk to see what approaches it had made towards blowing after so hot a summer', and received ten years later a request that he take 'some roots' with him to London to give his brother at South Lambeth (Bell, 1877, **2**: 158, where a note is given that the plant still flourished at Wakes in the 1870s).

41. Two, possibly three, sources exist for this awareness: first, through contact with his brother Benjamin's bookselling business he was kept abreast of what was current; second, because of guiding brother John's researches in Gibraltar, Pennant furnished him with some of the important Continental texts; third, his own reading of Stillingfleet had brought him into contact with the Linnaean developments.

42. This is very noticeable throughout the Gibraltar letters, White repeatedly letting John know that he has sent specimens to others for identification: for example, fish went to Pennant, insects to Forster (Foster, 1985a: 493).

43. Especially so in the *Journal*, where in the early years he added to the title-page: 'The Insects are named according to Linn: / The plants according to ye sexual system [Linnaeus in origin, but followed by Miller, and Hudson]: / The birds according to Ray'.

44. A monitory example occurs in the *Journal* for 28 Nov. 1771:

The reed-sparrow [*Emberiza schoeniclus*, reed bunting], passer torquatus, forsaking the reeds, & waterside in the winter, roves about among the fields, & hedges. This bird which I sometimes saw, but never could procure [obtain in the hand, usually by shooting] 'til now, I mistook for the aberdavine [*Carduelis spinus*, siskin].

Traces of this error of identification were included in *Selborne* (see the start of Barrington 8). See also Chap. 12, n. 9. Corrected items can also be found in the *Kalendar* and *Flora* (see Plate 5).

45. For example, when writing to Pennant about the latter's *British Zoology*, he commented: '. . . there is a passage . . . which you will pardon me for objecting to, as I always thought it exceptionable . . . [and] a false fact' (B.L. Add. MS 35138, White to Pennant, 2 Jan. 1769).

46. Selborne MS, White to Barrington, 15 Jan. 1770.

47. 24 Aug. 1770 and 27 July 1770, respectively.

48. B.L. MS 35138, White to Pennant, 29 Oct. 1770.

CHAPTER TEN

1. White wrote to John at Gibraltar on 26 Oct. and his brother replied on 10 Jan. 1759: for part of this latter, see Holt-White (1901, **1**: 110–13).

2. B.L. Add. MS 35138, White to Pennant, 28 Nov. 1768.

3. To gain preferment in the eighteenth century was no easy matter and one of White's first moves (26 Dec. 1769) was to send his brother a list of livings (over 150 in all) that were in the gift of the Archbishop of Canterbury, twin brother to the Governor at Gibraltar.

4. Foster (1985a: 231).

5. White arranged for Jack to be educated at Holybourne, near Alton, and occasional letters include a brief note from the schoolmaster, R. Willis, as well as from Jack himself (see Plate 7a).

6. For example, in Nov. 1771 White sent details of the Ringmer living; later (Mar. 1772) he advised him about the necessary 'letters testimonial for institution', and often cautioned him 'to sit as loose as possible to all expectations [of preferment] for fear of a disappointment'.

7. Bell (1877, **2**: 278); White to Marsham, letter of 20 Mar. 1792.

8. Foster (1985a: 315–16).

9. Butcher's-broom, known also as knee holly or prickly petigrue, bears small flowers on the midribs of its cladodes (pieces of flattened stem that look like leaves): it is as stiff and prickly as holly, and branches used to be bound together as a brush to scour benches and kitchen utensils: the native form is *Ruscus aculeatus*.

10. Worldwide there are some 75 species of swallow, grouped in 17 genera, the popular distinctions 'swallow' and 'martin' having no scientific relevance (swifts, which for White were hirundines, namely, swallows, are assigned today a quite different place in classification systems and are closely related to hummingbirds). For a useful outline of swallow behaviour, see Tate (1986).

11. B.L. Add. MS 35138, White to Pennant, 12 July 1770.

12. Giovanni Antonio Scopoli (1723–88) published *Annus I Historico-Naturalis* at Leipzig (1769); White termed this volume Scopoli's 'ornithology', and refers to him several times in *Selborne*.

13. Foster (1985b: 10).

14. Letter 10 to Barrington, dated 1 Aug. 1771.

15. In fact, on one occasion White *asked* Barrington to show him a famous collection:

> The Collection of Taylor White . . . is often mentioned as curious in birds, &c: can't I be introduced when in town to see this musaeum of my namesake's? (B.L. Add. MS 31852, 12 Apr. 1770).

16. For John White's correspondence with Linnaeus, see Bell (1877, **2**: 67–94).

17. This is from the same 12 Apr. letter

quoted at Note 15 above; the question of White's own writing is treated in the next chapter.

18. The most outspoken evidence of White's opinion occurs in a letter to Pennant of 2 Jan. 1769 (B.L. Add. MS 35138):

> In your Letter of June 28th: 1768 I could but admire with how much frankness you acknowle[d]ged several mistakes in your zoology with respect to some birds of the Grallae order. Candor is a very essential part of a Naturalist. And this accomplishment our great countryman Mr: Ray possessed in an eminent degree: & that rendered him so excellent. The great Northern explorer of Nature [Linnaeus] is not gifted with that grace in so ample a manner: & therefore, in that respect at least, is not so great a man. He is tenacious of his opinions, & knows not well how to retract. In his preface to his last edition of his tome of fossils, tho' a certain consciousness that he may be in the wrong seems to hang about him; yet he persists to the last, & cries in an idle jingle of words, "Nomine sed non omine mutavi [I have changed some names but not all of them]." If a man was never to write on natural knowledge 'til he knew every thing he would never write at all: & therefore a readiness to acknowledge mistakes on due conviction is the only certain path to perfection.

Later, in a letter to John, White hints at one cause of error in Pennant's work: 'He is a good correspondent, & knows a great deal: his great defect is an hurry of spirits, which pushes him on, & will not sometimes permit him to consider matters so cool[l]y as he ought to do' (Foster, 1985a: 326).

19. Foster (1985a: 493).

20. For example, see Bell (1877, **2**: 38).

21. The standard text on lacewings is Aspöck *et al.* (1980).

22. Foster (1985a: 325).

23. Discussion of so few specimens should not suggest that the remainder are of little interest; the letters are crammed with response to all kinds of birds, insects and fish, and offer absorbing study. They also include comment on the salting and labelling of specimens.

24. Foster (1985a: 326).

25. See Foster (1985a: 499–500, 495, 326) for this and the two preceding quotations.

26. Gifts to the Royal Society were acknowledged in *Philosophical Transactions*, and White's appears in Vol. 62 (1772).

27. Bell (1877, **2**: 69).

28. The main text for John's programme is White's letter of 10 Oct. 1769 (Foster, 1985a: 231–2), and it is from that letter that the principal data for discussion are derived.

29. 'He [Pennant] publishes y^r fishes in Gent: mag: thro' a pruriency for publishing which he cannot suppress', wrote White to John in 1772 (Foster, 1985a: 498); in fact Pennant used three items sent by John to White to launch a series of articles, with plates, about 'inhabitants of air, earth, or water, which have been slightly noticed or quite overlooked by other writers' (*The Gentleman's Magazine*, **41**, June 1771: 249).

30. John's letters to Linnaeus are necessarily formally academic, but the only part of his work to survive (*The Introduction to Fauna Calpensis*, ed. W. H. Mullens, 1913) is an attractive piece of writing and includes several plans and views of Gibraltar drawn by John himself.

31. Bell (1877, **2**: 99).

32. A useful measure of contemporary understanding is given in remarks printed in *The Gentleman's Magazine* (1785: 855) which give details about the founding of a society for promoting the study of natural history: 'the several branches of Natural History which the Society has undertaken to promote . . . are zoology, botany, conchology, entomology, mineralogy, and extraneous fossils'. For a modern account of the establishment of one specialised discipline, see Farber (1982).

33. Foster (1985a: 493).

34. The text White refers to by James Lee (1715–95), the famous nurseryman and gardener responsible for the intro-

duction of the fuchsia, was a translation of Linnaeus's explanation of the sexual system. Lee first published his volume in 1760, and there were several later editions: see Willson (1961).

35. The substance of these proposals is given in Foster (1985a: 318, 321–2).

36. The idea of an introduction composed of 'the civil history of Gibraltar from ye earliest times' occurs in White's letter to John of 30 Aug. 1770 (Foster, 1985a: 318); he was as good as his word, and wrote on 8 Oct. 1770 to Barrington:

> As I know you are a Gentleman of universal reading, I have a great favour to ask of you, which yet I trust you will not deny me; & that is that you would please to furnish me with some anecdotes concerning the anti-quities & civil history of Gibraltar. Some of the old greek geographers may perhaps mention how that spot was circumstanced in old times. So particular a situation must have been known to the Phoenicians, Greeks, Romans, & Saracens. The famous pillars of Hercules are supposed to have stood some where near where our garrison stands (Selborne MS).

37. For an understanding earlier in the century of the '*Five* sundry Streams, which pass thro' [the Straits] out of the *Atlantick* Ocean, into the *Mediterranean*', see Barlow (1722: 129–88).

38. Stillingfleet (1762: 282); an entry given by Berger on 4 Aug. for the calendar at Uppsala.

CHAPTER ELEVEN

1. B.L. Add. MS 31852, White to Barrington, 12 Apr. 1770.

2. Foster (1985a: 325).

3. Scopoli (see Chap. 10, n. 12) wrote about Carniola, in his time a province of the Austrian empire which today is the western region of Slovenia, Yugoslavia.

4. B.L. Add. MS 35138, White to Pennant, 14 Sept. 1770.

5. Although these occur later than the comment to Pennant, they are important because they link together White's reading of Scopoli and his adherence to local study:

> The anni of Scopoli [he wrote to Barrington on 8 Oct. 1770 from Ringmer] are now in my possession; & I have read the annus primus, the ornithology of Carniola, with some satisfaction ... men that undertake only one district are much more likely to advance natural knowledge than those that grasp at more than they can possibly be acquainted with (Selborne MS).

Similar views were sent to John when he was at Gibraltar. Apparently John was planning to extend his enquiries beyond the Rock and into Spain, and White commented:

> The more confined y^r sphere of observation is the more perfect will be y^r remarks: & it seems as if y^r rock afforded you matter sufficient. However I will mention y^r proposal to Mr Barrington, Dr Bosworth &c: & see if any encouragement to look more abroad can be met with ... True naturalists will thank you more for the life & conversation of a few animals well studied & investigated; than for a long barren list of half the Fauna of the globe. Monographers are the best Nat: Historians (23 Oct. 1771; Foster, 1985a: 491, 492–3).

6. Brief comment on two of these collections was made in a letter to Pennant on 1 Aug. 1770 (B.L. Add. MS 35138):

> When I was last in town our friend Mr: Barrington most obligingly carryed me to see many curious sights ... [and] many astonishing collections of stuffed & living birds from all quarters of the world.

7. Foster (1985a: 320).

8. B.L. Add. MS 31852, White to

Barrington, 12 Feb. 1771. The passage continues:

> As I never dreamed 'til very lately of composing any thing of this sort for the public inspection, I enter on this business with great diffidence, suspecting that my observations will be deemed too minute & trifling. However if I ever finish it I shall submit it to yr better judgment.

9. Foster (1985a: 324); letter of 25 Jan. 1771.

10. *Selborne*, Barrington 66.

11. Foster (1985a: 494). Yet, by the time White decided to write about Selborne little help was forthcoming from either Skinner or Sheffield: 'As to Skinner he is so lazy that he will not write one common friendly letter in a twelvemonth: & Sheffield is so busy that we must expect little help from that quarter' (Harvard bMS Eng 731–131, White to John White, 8 May 1775). The activities of Sheffield were focused at this time on whether he would become head of his college (Barrett, 1987).

12. It was in preparation for this journey that Pennant published his list of 'Queries': see Appendix F.

13. In July, early in their tour, they called on Skinner, who was then on a visit in Wales (at Brecon), and it was from him that White received some account of their progress up to that time (Bell, 1877, **2**: 234–235).

14. Two of Lightfoot's letters to White are given in Bell (1877, **2**: 231–4); for an evaluation of *Flora Scotica* and some details of the author, see Slack (1986).

15. Selborne MS, White to Pennant, 2 Sept. 1774.

16. White to John, 2 Nov. 1773: see Bell (1877, **2**: 18–19).

17. Bell (1877, **2**: 22); the letter is dated 12 Jan. 1774. For earlier mention of White's agreement to write these monographs, see (again in Bell) White's letter to John of 9 Dec. 1773.

18. For White's travel dates, see *Journal*; and for some account of what occurred during the visit to London, see a post-script (written in London) to a letter begun on 6 Mar. 1774, and one from Selborne (on return from London) dated 29 Mar. 1774, both of which are addressed to John (Bell, 1877, **2**: 26–8).

19. White to John, 29 Mar. 1774 (Bell, 1877, **2**: 27).

20. And so also were specimens retained: 'All y.ᵉ birds, &c: are preserved with care: so that when ever you return you may (as will be very needful) re-examine them with more care & exactness' (Foster, 1985a: 499, 500; letter of 9 Mar. 1772).

21. For this and the following quotation, see respectively letters dated 11 Sept. 1773 and 12 Jan. 1774 in Bell (1877, **2**: 15–16, 22).

22. See *Selborne*, Pennant 10.

23. Bell (1877, **2**: 28). Although it is clear that Barrington had an important role in bringing White to this decision, it was Pennant who seems first to have suggested that White publish in 'a periodical publication, that shall receive the various pieces of natural history that otherwise might perish': White acknowledged that he was 'ready to advance [his] mite' (this was in 1769), but the idea seems to have progressed no further (B.L. Add. MS 35138, White to Pennant, 1 Sept. 1769).

24. Mulso (1907: 266).

25. That brother John was progressing well with his 'Fauna' must also be taken into consideration. 'Your Fauna,' wrote White to John in Oct. 1775, 'to which I think myself at least a foster-father, is become, I hear with pleasure, a fine thriving child. I could be glad to examine its features, and to dandle it, and remark how it shoots up towards its ἡλικια [full flowering]' (Holt-White, 1901, **1**: 291).

26. Thomas Barker and his son, Samuel, visited Selborne in autumn 1775, and in Nov. White wrote to them: 'I desire that both of you would send me every hint in Nat. Hist. that occurs to your minds after your recent visit to these parts' (Bell, 1877, **2**: 115); and Mulso had been shown the work in the summer, after which he wrote to White:

You have a double Felicity in your Manner of Entertainment; You can gratify your Visitors both wth beautifull Originals, & high Descriptions; Representations studiously copied from Nature & finished with a Masterly Hand. As You intend your Works for ye Public, I would not say so much in a Strain of Flattery . . . You have happily grounded Ethics on a stable & beautifull Basis, ye Works of God . . . This is my real Opinion of your Work (Mulso, 1907: 258),

27. In *Selborne*, the continuous sequence of real letters sent to Barrington ends with Barrington 22 dated 13 Sept. 1774; although some of the later letters are also based on real correspondence, the very next letter, Barrington 23 (dated 8 June 1775), is derived from a communication of 8 Apr. 1775 that White sent to his brother John at Blackburn (Harvard 731/130).

28. Bell (1877, **2**: 128). White's own inspection of Grimm's work took place in Jan. 1776, and he conveyed his good opinion to brother John in May the same year (Bell, 1877, **2**: 48, 53–4), which is an interesting decision in the light of the contradictory advice he had received (Holt-White, 1901, **1**: 288–9). For a modern view, see Clay (1941).

29. Bell (1877, **2**: 120).

30. 'Antiquities', Letter 24. For an excellent recent history of the Priory, see Le Faye (1975: 47–71).

31. For this and the preceding quotation, see Bell (1877, **2**: 63, 132).

32. Bell (1877, **2**: 137); letter of 2 Sept. 1778 to his nephew, Samuel Barker. By this time work on the antiquities had been in progress for several years and White found some of the results highly instructive. To his brother Thomas he remarked that, although 'the lands of the convent and the Templars abutted on each other, and were intermixed, yet we see that those two societies of Religious lived on the best of terms, in an intercourse of mutual good offices, exchanging lands and permitting roads to be opened for each other's mutual convenience' (Bell, 1877, **2**: 131–2): which is an enlightening comment from a man living in the 1770s. And from one of the Bishop's Registers he had discovered that the ecclesiastics at the Priory.

> had deviated much from its original simplicity . . . had become mighty hunters . . . used to let *suspectae* come into their cloisters after it was dark . . . had administered the sacrament with such nasty cups and such nasty sour wine that men abhorred the sight . . . [and that they] were got into a method of lying naked in bed without their breeches (Bell, 1877, **2**: 136).

33. Richard Chandler (1738–1810), classical scholar and antiquary, specialised in remains from Greece and Asia Minor; in 1772 he was Senior Proctor at Oxford, and a fellow at Magdalen, a college that later presented him to livings near Selborne. For White's reports about his work at this time, see Bell (1877, **2**: 187ff).

34. Letter of 2 June 1782 (Mulso, 1907: 306); see also letter of 21 Sept. 1780, where Mulso reproved White in similar terms:

> Pray does your Book come out this Winter? I really cannot hold out any longer. If You spoil the genuine Elegance & neat Simplicity of the original Design, by a Farrago of Antiquities, routed out of the Rusts & Crusts & Frusts of Time, I shall not esteem it so well as I once did; & so I tell You.

The date of these quotations is of interest: Mulso moved on the fringes of the court establishment, and evidently an appreciation of the Gothic was not prevalent.

35. Richard Chandler has already been mentioned, but equally important were Ralph Churton (1754–1831), a native of Cheshire, later fellow of Brasenose, Oxford, and John Loveday of Caversham (see Markham, 1984).

36. Most notable of these pursuits were the discoveries made about the antiquities of the parish. The principal in this activity appears to have been brother Thomas, who 'opened two of the most

promising tumuli' on Selborne Down (see *Journal* at 15 Sept. 1782), but White himself had initiated enquiries about local remains much earlier: see, for example, a letter to White of 1777 from William Sewell, rector at the neighbouring village of Headley, about the Roman medals which had been found earlier in the 1770s in Wolmer Pond (Bell, 1877, **2**: 393–4), and for evidence of an earlier discovery see *Selborne*, Pennant 8. For details about the loss of some remains, see Letter 26 of 'The Antiquities'.

37. For the hermitage, see Bell (1877, **2**: 126); the farmland became available early in Aug. 1777, and White wrote immediately to brother Thomas since 'I shall now, I trust, be able to secure that "*Angulus*" which the family have been labouring so long to obtain' (Holt-White, 1901, **2**: 14); as for the Bostal, the *Journal* records for 27 Sept. 1780: 'Finished a *Bostal* or sloping path up the hanger from the foot of the zigzag to the corner of the Wadden . . . A fine romantic walk, shady & beautiful'.

38. This arose mainly because Thomas inherited at this time monies from the death in 1746 of Thomas Holt, receiver to the Duke of Bedford, whose affairs White himself attended to as a young man at Thorney, Isle of Ely. It was at this time (1776–7) that Thomas conceived the idea of preparing an account of the natural history and antiquities of Hampshire: he was elected a Fellow of the Royal Society in Jan. 1777. For details of litigation in connection with the inheritance, see Holt-White (1901, **1**: 297–314), which also gives in a letter from White to Thomas of 4 Jan. 1776 an important account of White's advice to Thomas concerning the projected history of the county.

39. Bell (1877, **2**: 137); letter from White to Mary White 30 Sept. 1780, which also reveals the friendly rivalry between the advocates of the new path through the hanger and those who supported the merits of the Zigzag.

40. The relationships between the Whites and the Yaldens is complex. White himself had been curate at Newton Valence (and West Dene) for Reverend Edmund Yalden: as vicar of Newton Valence Edmund was followed by his son, Richard, on whose death in 1784 the living passed to Edmund White, a son of White's brother Benjamin. Richard Yalden's brother, William, was a general merchant in London, and took into partnership White's brother Thomas: on the death of his partner, Thomas married his partner's widow, Mary, in 1758 and they had three children: Mary (Molly), who helped with 'The Antiquities', and the twins, Thomas and Henry. Further, White's brother Benjamin married first Anne Yalden, sister to Richard Yalden, and later (1786) Richard Yalden's widow (who was also called Mary).

41. For the history of Wakes as a building, see Meirion-Jones (1983: 145–69).

42. Bell (1877, **2**: 12); letter of 26 June 1773; see also letters in the same volume to John dated 6 Mar. and 29 Apr. 1774, and 9 Mar. 1775.

43. Mulso (1907: 248); letter of 15 Feb. 1774.

44. Details of some of the costs are given in Holt-White (1901, **2**: 50–1): the looking-glass and the 'flock sattin' wallpaper cost just under £10 each; the carpet, 'a fine stout large Turkey', was eleven guineas.

45. Holt-White (1901, **2**: 52). The only problem with the room seemed to be the chimney, since there is a note in the *Journal* at 1 Jan. 1788:

> Contracted my great parlor chimney by placing stone-jams on the top of the grate on each side, & building brick-work on the jams as high as the work-man could reach. This expedient has entirely cured the smoking, & given the chimney a draught equal to that in the old parlor.

46. White was curate at Faringdon until 1784, when he accepted the curacy at Selborne consequent upon the death of his close friend the Reverend Andrew Etty; the new vicar, Charles Taylor, was non-resident, although he chose to be married at Selborne in 1786 (Holt-White, 1901, **2**: 153). It is also worth noting White's continued, and meticulous,

administration of several charities established by his grandfather.

47. Mrs Snooke died on 8 Mar. 1780, and Timothy then came to Wakes: she (for that was the sex, determined after death in 1794) provided considerable distraction and many of the entries in the *Journal* reveal White's avid observations (Townsend Warner, 1946); and not only scientific ones but psychological and interpretative as well—see Bell (1877, **2**: 182–5), who gives the letter White wrote in the persona of the tortoise which he addressed to a daughter of Mulso's after the family had stayed with him in Aug. 1784.

48. This and the preceding quotations are part of White's early invitations to Ralph Churton: see Bell (1877, **2**: 186–92).

49. Although the Jacobite threat of mid-century had come to little, in the decades after the successes of the Seven Years War (1756–63) Britain became increasingly anxious: the Boston Tea-party (1773) and the American Declaration of Independence (1776) were followed by defeat at Saratoga (1777), declaration of war by France (1778) and Spain (1779), and surrender at Yorktown (1781); this last White entered in the *Journal* (19 Oct.) and termed the day 'ill-fated'.

50. Richard White (1761–1816) was one of brother Benjamin's sons. In Aug. 1776 White wrote to his nephew, Samuel Barker, 'Nep. Richard, who has left school, is here with me', and followed this in a letter of Nov. 1776 to brother John at Blackburn with:

> D— is with me; he is good natured, but somewhat heady at times. It is well he is intended for trade, since he loves anything better than [a] book: bodily labour he does not spare; for rolling, wheeling, water-drawing, grass-walk sweeping are his delight. I have taught him to ride; and perhaps a good seat on an horse may be more useful to him than Virgil or Horace. I tryed Phaedrus, but my patience fail'd. However, he may procure

health and strength and a little behaviour at my house (Bell, 1877, **2**: 129 and 58 respectively, and see also p. 139).

51. Samuel Barker (1757–1835) appears to have written to White first in 1770, and he received instruction over the years in most of his uncle's interests, including the identification of plants. It is from this latter field that we know particularly of White's association with William Curtis, author of the famous *Flora Londinensis* (1775–98): see Bell (1877, **2**: 124), which tells of Curtis sending help to Samuel via White on the identification of mosses. For details of how the Whites, especially Thomas, assisted Curtis to establish the London Botanic Garden (1779), see H. Curtis (1941).

52. Many of these letters are given in Bell (1877), but a long sequence, presently at Fitzwilliam Museum, Cambridge, awaits modern editing.

53. It is from a letter to Edmund shortly before this (21 Feb. 1783) that we learn on a serious note of White's private devotions and of his manner in church:

> I could wish that you would make it a rule to read aloud to yourself every day some portion of S[cripture] or the [Book of] Common Prayer, tho' ever so short, and that you would also sometimes read before a judicious friend—but at the same time plainly and unaffectedly; and do not aim at anything theatrical or fine, but only attend well to what you read; and your own good sense and ear will tell you at the time how to modulate your voice, and lay your accents justly according as you are affected by what is before you (Bell, 1877, **2**: 150–1).

54. Bell (1877, **2**: 193); letter to Ralph Churton, 4 Jan. 1783.

55. The most appealing count of nephews and nieces (at the time they had increased to 56, which included those related by marriage as well as blood relations) is associated in a *Journal* side-note with the date 16 May 1790: 'One polyanth-stalk produced 47 pips or

blossoms.' The last *Journal* entry to record an addition to the total was made on 11 Jan. 1793 (Louisa Neave, bride to Gibraltar Jack) and brought the number to 62.

56. Bell (1877, **2**: 196); letter to Ralph Churton, 20 Aug. 1783. Several of White's poems are dated 1782–3, and are given in Bell (1877): 'On the Rainbow' has the date 13 Feb. 1783, and White says in the cited letter to Churton that 'A Harvest-Scene' was written during autumn 1782.

57. Although Bramshott was so near (seven miles in a direct line through Wolmer Forest), from 1777 White and the Richardsons often exchanged an annual visit, staying at their respective homes for one or two nights. The *Journal* shows that White took a keen interest in the different conditions in the garden at Bramshott Place compared with his own at Selborne, an entry for 3 July 1783 being representative: 'Mr Richardson's garden abounds with fruit, which ripens a fortnight before mine. His kitchen-crops are good, tho' the soil is so light & sandy. Sandy soil much better for garden-crops than chalky'.

58. Jack's apprenticeship was over on 16 June 1781 and he arrived with his mother at Selborne on 31 July that year (Holt-White, 1901). For some account of his training, see Cash (1963: 273–5).

59. Not the modern rhubarb, *Rheum rhaponticum*, but *Rumex alpinus*, which, together with *R. patientia*, was grown in White's time for culinary use. The stand of *R. alpinus* had arrived the same year from London (*Journal*, 20 Mar.) and later there are the notes: 'Monks rhubarb seven feet high; makes a noble appearance in bloom' (1 June 1789) and 'Cut the leaves of Rhubarb for tarts: the tarts are very good' (2 May 1792).

60. For a full listing of all White's truffle records together with a short introduction, see Foster (1986a: 42–7). One of the late records gives White's culinary judgement: 'Stewed some trufles: the flavour of their juice very fine, but the roots hard, & gritty. They were boiled in water, then sliced, & stewed in gravy' (*Journal*, 4 Nov. 1790).

61. Considerable discussion took place in 1775–6 about the title of John's work (Bell, 1877, **2**: 49–51): 'Calpensis' was chosen because it referred to the Latin name for the more northerly of the two Pillars of Hercules, the Rock of Gibraltar.

62. The most famous example of this concerns the engendering of earthworms which is reported in *Selborne*, Barrington 35 dated 20 May 1777, but there are several later examples.

63. Bell (1877, **2**: 168); letter to Samuel Barker.

64. Writing to Ralph Churton on 4 Aug. 1788, White expressed some surprise at the cost of the engravings: 'My brother and nephew have spared no expense about [my book], and particularly on the engravings, which have cost a considerable sum', and continued: 'This book will . . . not be published now till the autumn, when the town begins to fill' (Holt-White, 1901, **2**: 185).

65. Quoted in Holt-White (1901, **2**: 187): the entry is dated 3 Dec. 1788. Although Henry could have used the format of Barrington's *Journal*, he chose to make up his own. A complete run of these volumes, which are more social than scientific in content, appears to be no longer extant. A brief discussion of those that do exist (at the British Library and at the Bodleian) is given in a booklet (Burton, 1979).

66. 'The Topographer for the year 1789', vol. 1, part 1, as quoted by Holt-White (1901, **2**: 194–5). The review in *The Gentleman's Magazine* (1789: 60–3, 144–6) was by brother Thomas, and is appropriately discreet.

67. See *Journal* 2 Oct. 1790; and for *P. puffinus* 19 Sept. 1790. Both birds came from Lord Stawell, and White recommended that 'Elmer of Farnham, the famous game-painter' should take a copy of the bird: this was done, and a painting was sent to White; it was published in *A Naturalist's Calendar* (1795) and in several later editions of *Selborne*.

68. To appreciate fully White's use of poetic extracts one needs to meet them in context, embedded in the data of the daily record, but these two examples may give

an indication of his mood:

21 June 1790:
Scarlet straw-berries good. A small praecox melon.

The longest day – – –
The longest daye in time resignes to nighte;
The greatest oke in time to duste doth turne;
The Raven dies; the Egle failes of flighte;
The Phoenix rare in time herselfe doth burne;
The princelie stagge at lengthe his race doth ronne;
And all must ende that ever was begonne.

[Geoffrey Whitney, *Emblems* (1586)]

8 July 1792:

The Poet of Nature [James Thomson, author of *The Seasons*] lets few rural incidents escape him. In his *Summer* he mentions the *whetting* of a *scythe* as a pleasing circumstance, not from the real sound, which is harsh, grating, & unmusical; but from the train of summer ideas, which it raises in the imagination. No one who loves his garden & lawn but rejoices to hear the sound of the mower in an early, dewy morning.

Echo no more returns the *chearful* sound
Of sharpening scythe – – –

69. Letter to Ralph Churton, 1 Sept. 1789, in Bell (1877, **2**: 213).

70. White to Marsham, 20 Nov. 1792; see also White to Churton, 26 Jan. 1793 (Bell, 1877, **2**: 293–4 and 228–9 respectively).

71. Robert Marsham (1708–97), much of whose correspondence together with White's replies is given in Bell (1877), was not the only stranger to write: the previous year George Montagu (1751–1815), the author of the future *Ornithological Dictionary* (1802), wrote—and there must have been others.

72. It was Marsham who advised White that the tree known to local residents as the Grindstone Oak in Holt Forest was supposed the largest in Britain, and (at seven feet from the ground measuring 34 feet in circumference) should be included in a second edition of *Selborne* (Bell, 1877, **2**: 247). That White omitted specific mention of the specimen is somewhat surprising since brother Thomas had established a reputation as an arboriculturist through a series of articles in *The Gentleman's Magazine*: see for example vol. 55 (1785: 109–12, 598–600), where he writes on the oak and the ash. Execution of the idea of publishing an engraving of the Grindstone Oak was overseen by White's publishing brother, Benjamin, who had now retired from London to live at Marelands, an estate at Bentley near Holt Forest that he rented from Lord Stawell (Bell, 1877, **2**: 302).

73. In particular, the two legends attached to the vignette of the hermit's cell which serves as a preface to the natural history part of the volume. The one is from Homer: 'It is a rough country, but a good place for rearing young men. I for my part cannot envisage anything sweeter than one's own country'; the other is from Cicero: 'In short the whole of that territory of ours is rough and mountainous, [but] dependable, trustworthy and a protector of her own people'.

CHAPTER TWELVE

1. 'Should I ever be able to finish my work respecting this my native place, the old deeds and charters &c. will furnish a large appendix': White to Samuel Barker, 2 Sept. 1778 (Bell, 1877, **2**: 137).

2. Selborne MS, White autograph dated 15 Jan. 1770 to Barrington.

3. This is an important factor in the continued appeal of White's work for British readers; the *level* of discussion is another, but of universal relevance.

4. The manuscript, with assistance from public subscription, was purchased in 1980 by the Trustees of the Oates and

Gilbert White Museum at Selborne; the curator, Dr June Chatfield, has kindly permitted access to the MS prior to publication of a facsimile edition.

5. The phrase is Jack Stillinger's, who accomplished in 'Gilbert White to Thomas Pennant: Two Original Letters at Harvard' (*Harvard Library Bulletin*, 1957) the only significant work on White manuscript sources from the 1930s until recently.

6. This is indicated by the Selborne MS versions of these letters, which are in the hand of Gibraltar Jack, who left Selborne for Lancashire and a medical apprenticeship in November 1774: for this last, see Bell (1877, **2**: 107).

7. Barrington 25 and following.

8. In a letter to Samuel Barker of 15 Nov. 1775 White asked that both Samuel and his father, Thomas Barker, send him any suggestions they might have consequent upon their visit to Selborne that year (Bell, 1877, **2**: 115). The request was repeated a few months later. As for results of these requests, Barrington 30 dated in *Selborne* 3 Apr. 1776, which describes dissection of a cuckoo and a fern-owl to establish whether bone structure inhibited female incubation, is associated with an enquiry of Samuel's; and so also may be Barrington 40 (2 June 1778), which considers vegetative husbandry — see respectively letters of 6 Sept. 1775, and 19 Aug. 1776 (Bell, 1877, **2**: 112–114, 127).

9. Some indication of the decisions that had to be made in London by relatives is given by: 'Might not the Hermitage print come in well at the back of the first title-page, or as a tail-piece to the "Natural History"?' (White letter of Feb. 1788, given in Bell, 1877, **2**: 171); and 'As to the Antiq. letters being addressed to some body, I cannot well tell what to say: though I think if to any, it should be to D^r. Chandler; . . . But . . . I think it may be best to let the letters go as they are, addressed to *no one*' (Holt-White, 1901, **2**: 182).

A fascinating example of inept collaboration which contributed to the 'charm' of *Selborne* is provided by the beginning of Barrington 8 (dated 20 Dec. 1770). The letter begins with 'The birds that I took for *aberdavines* were reed-sparrows (*passeres torquati*)', and has been considered something of a puzzle since there is no prior mention of aberdavines (*Carduelis spinus*, siskin) in the *Selborne* letters to Barrington. The sequence of decision underlying the inclusion of the sentence is complex, but briefly: real letters of 12 Apr. 1770 and 20 Dec. 1770 to Barrington include mention of aberdavines, but late in 1771 White realised his identification was in error and entered in the *Journal* on 28 Nov.:

> The reed-sparrow [*Emberiza schoeniclus*, reed bunting], passer torquatus, forsaking the reeds, & waterside in the winter, roves about among the fields, & hedges. This bird which I sometimes saw, but never could procure 'til now, I mistook for the aberdavine.

Consequent upon this discovery he then made a series of corrections to entries in the *Journal* (23 Jan., 17 Mar., 29 Mar., 17 Apr. 1770 for example), and added to his original letter to Barrington of 20 Dec. 1770 the sentence cited. Finally, when editing the letter for publication, several paragraphs of the letter, including discussion of aberdavines, were marked as not to be set by the printer. However, the inserted sentence must have seemed to be a corrected form of what was to be omitted and was included. That this result was not White's intention is shown, I think, by comparing this outcome with his treatment of the real letter of 12 Apr.: in this instance White made no correction to the letter, and when editing marked the entire discussion for deletion; it is not included in *Selborne*.

10. Plot, for example, organised *Oxfordshire* into chapters the sequence of which was planned 'to observe the most natural method that may be', and this resulted in: 1. 'Of the Heavens and Air'; 2. 'Of the Waters'; 3. 'Of the Earths'; 4. 'Of Stones'; 5. 'Of Formed Stones [fossils]'; 6. 'Of Plants'; 7. 'Of Brutes'; 8. 'Of Men and Women'; 9. 'Of Arts [things made by

man: roads, carts, meteorological instruments, husbandry . . .]'; 10. 'Of Antiquities'.

11. An interesting indication of why the private scientific letter gave way to the establishment of scientific periodicals and scientific societies occurs in two articles by McKie (1972).

12. For a lucid discussion of the form in eighteenth-century literature, see Todd (1966); Anderson *et al.* (1968) is also pertinent.

13. *The Storm* (1704), which gives an account of a tempest in Nov. 1703, is a good example of Defoe's semi-fictional writing, in which some letters are genuine and others fabricated.

14. The Royal Academy of Arts for example was founded in 1768, the Linnean Society in 1788.

15. Plot (1640–96) is an interesting figure in the history of science: he was a Professor of Chemistry at Oxford, the first Keeper of the Ashmolean, and Secretary of the Royal Society. For his approach to the gathering of materials for a general natural history, see Günther (1939).

16. This did not, however, prevent his attending occasional meetings when he was in London (although, as Sheila Rainey reminds me, he was anxious about visiting at times of disturbance, for example, during the Gordon Riots), and recommending Ralph Churton if he was

> in London on a Thursday . . . to attend on the R.S. and Antiquary-meetings in their new splendid rooms at Somerset-house. Dr. Chandler can probably put you in a method of being introduced; if you do not see him, attend in the outer room . . . and enquire for Dr. Lort, who, I trust, on your using my name will introduce you to both the meetings, where perhaps you may hear somewhat worth your trouble (White to Churton, 30 Mar. 1784, Bell, 1877, **2**: 198).

17. A detailed literary study of White's use of the letter form (a volume much needed) would have to tackle the varied action—that of the narration, of White's

view of the narration and of the reader's response as issues and ideas are grasped and contributed to from prior experience—which present the village and its natural history as so many Chinese boxes, all contributing a fresh layer of experience to a whole that appears to be diminishing but which is also growing. A modern equivalent of what *Selborne* is in scale might be a television documentary, where the results of many hundreds of hours of filming and editing underlie a few minutes of finished product: the outcome is more a contribution to art than to science (although science can certainly be served by such means). For some discussion of this approach to natural history, albeit not to *Selborne*, consult David Attenborough, Huw Wheldon Lecture, 1987.

18. It all depends on what one calls a 'distinct discovery' but these are most commonly cited: differentiation of the three English species of leaf warbler; discovery of the harvest mouse and the noctule bat; and identification of the lesser whitethroat.

19. There are various ways of enforcing such a judgement, but it is unusual, for example, to find any mention of White in standard histories of biology; moreover, at no time in his work does he *discuss* either of the two dominant issues of the day—the geologic age of the Earth, and the nature of species. Evidently, such discussion was not to his taste. As he remarks in a letter to Pennant (Pennant 24), who had enquired whether White had any views concerning the presence in America of species unknown in Europe:

> Ingenious men will readily advance plausible arguments to support whatever theory they shall chuse to maintain; but then the misfortune is, every one's hypothesis is each as good as another's, since they are all founded on conjecture.

And yet, awareness of the issues can be seen in many of the *Selborne* letters and it clearly informs his attitudes, principally through reading Plot and Ray. Readers interested in pursuing this topic may care to consider the incluison in *Selborne* of, say, the fossil plate, and the account of the Goodwood moose, a creature at the time thought by some to indicate that the Irish elk was not extinct (Gould, 1978; Rolfe, 1983: 263–90). Glass (1959) is a useful background text to these issues.

20. And much more: that earthworms are hermaphrodites, notions of bird territory, ideas about coloration and sexual selection, pursuit of the relation between climate and natural events, comparison of local climates, discovery that the ring ouzel, *Turdus torquatus*, is a passage migrant in southern England are some of them. But they are all individual, discrete, and in that sense representative of the world of the first half of the century. The leading edge of what was happening later was well put by Buffon as early as 1744:

> It is not necessary to imagine . . . that, in the study of natural history, one ought to limit oneself solely to the making of exact descriptions and the ascertaining of particular facts. This is, in truth, . . . the essential end which ought to be proposed at the outset. But we must try to raise ourselves to something greater and still more worthy of our efforts, namely: the combination of observations, the generalization of facts, linking them together by the power of analogies, and the effort to arrive at a high degree of knowledge. From this level we can judge that particular effects depend upon more general ones; we can compare nature with herself in her vast operations (Lyon and Sloan, 1981: 19).

Thus might Baconian ant-and-spider-science be replaced by bee-science (see Ogborn, 1985: 10–12).

21. Of these two quotations the first is from Gerard Manley Hopkins, the second from John Keats. Both these poets belong to the Romantic tradition and in that sense open up connotations that are inappropriate for White, especially in their experience of the aesthetic appeal of Nature; but as fragments of poetic utterance White would know what the words meant. Nevertheless, White's own relationship to Nature was ultimately more a matter of labour and reason. In a sermon (Bell, 1877, **2**: 308–15) that he preached regularly over 35 years on the parable of the talents, he stressed men's 'capacity (if we will but take care rightly to apply [our talents]) of doing service to God, and men'. Of all the talents, he esteemed most highly '*Reason* . . . the principle of all Knowledge, both of things in the world, and those above it', and hence was in full sympathy with John Ray's views in the latter's *The Wisdom of God Manifested in the Works of the Creation* (1691), although his quotations tend to be taken more frequently from the very similar *Physico-theology* (1713) of William Derham. Just occasionally, however, his writing hints at a particular phenomenon that seems to evade his reason (consider for example, when a viper was caught, the removal of fangs with scissors; or the near absence in the *Journal* record of magpie and jay, birds that destroy eggs and nestlings of other species), but his faith in the beneficence of Providence is never thereby threatened. This faith is in fact so strong that it informs everything he wrote, including the meteorological record where his gratitude is often acknowledged through noting the 'sweetness' of the weather: see Chap. 9, page 101.

APPENDICES

APPENDIX A

The White Family

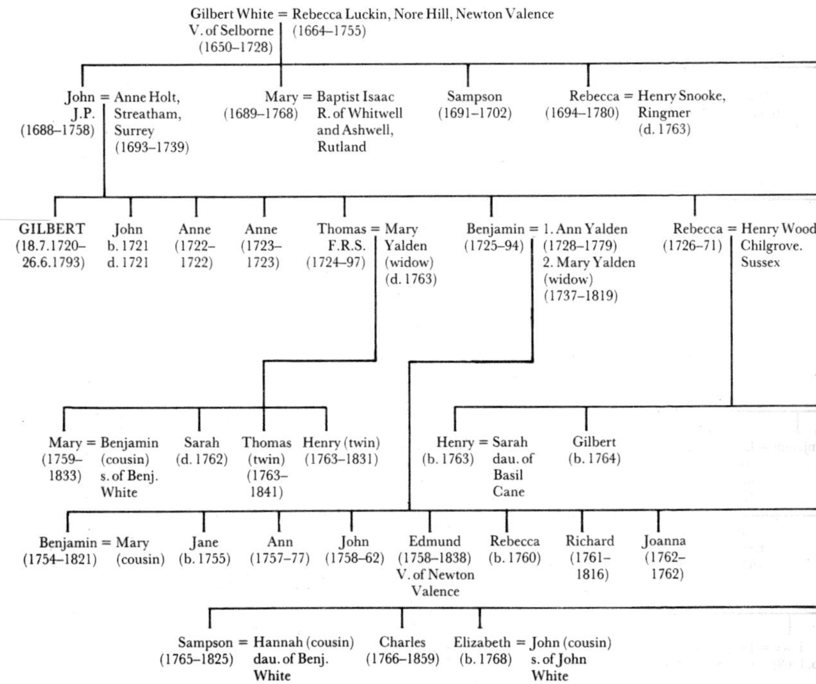

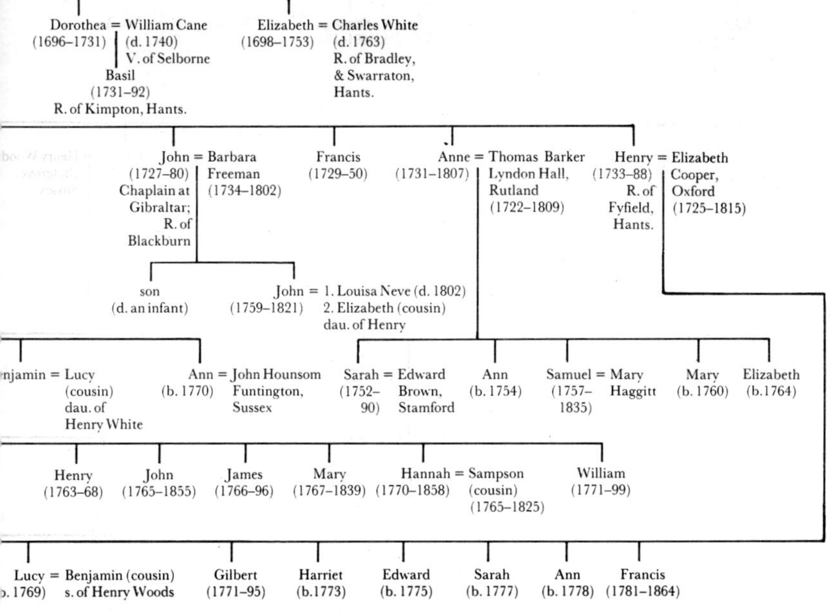

Dorothea = William Cane
(1696–1731)   (d. 1740)
V. of Selborne
Basil
(1731–92)
R. of Kimpton, Hants.

Elizabeth = Charles White
(1698–1753)   (d. 1763)
R. of Bradley,
& Swarraton,
Hants.

John = Barbara
(1727–80)   Freeman
Chaplain at   (1734–1802)
Gibraltar;
R. of
Blackburn

Francis
(1729–50)

Anne = Thomas Barker
(1731–1807)   Lyndon Hall,
Rutland
(1722–1809)

Henry = Elizabeth
(1733–88)   Cooper,
R. of   Oxford
Fyfield,   (1725–1815)
Hants.

son
(d. an infant)

John = 1. Louisa Neve (d. 1802)
(1759–1821)   2. Elizabeth (cousin)
dau. of Henry

njamin = Lucy
(cousin)
dau. of
Henry White

Ann = John Hounsom
(b. 1770)   Funtington,
Sussex

Sarah = Edward
(1752–   Brown,
90)   Stamford

Ann
(b. 1754)

Samuel = Mary
(1757–   Haggitt
1835)

Mary
(b. 1760)

Elizabeth
(b.1764)

Henry
(1763–68)

John
(1765–1855)

James
(1766–96)

Mary
(1767–1839)

Hannah = Sampson
(1770–1858)   (cousin)
(1765–1825)

William
(1771–99)

Lucy = Benjamin (cousin)
b. 1769)   s. of Henry Woods

Gilbert
(1771–95)

Harriet
(b.1773)

Edward
(b. 1775)

Sarah
(b. 1777)

Ann
(b. 1778)

Francis
(1781–1864)

White's Extended Garden

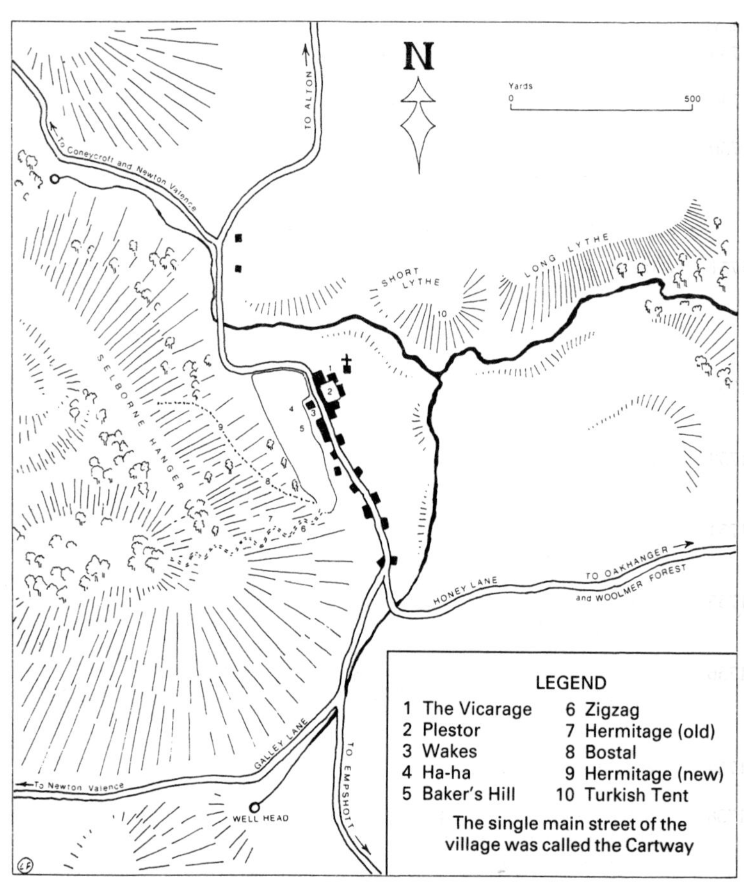

APPENDIX C

Development of the Garden at Wakes

YEAR	ITEM	REFERENCE
1730	White's planting of oak and ash	NJ (Aug. 1790)
1740	White's father lays out walks in the outlet	Bell, **2**: 259
1750	Arbour on hill White plants elm Henry Pelham (Esher Place) builds belvedere for White to see	JM 34 NJ (Aug. 1790) JM 38

1751	White begins *Garden-Kalendar*; plants
	   quincunx of firs on Baker's Hill, and
	   shrubbery
	'Turkish' tent mentioned
		GK Oct.
		
		JM 55

1752	Bench in field mentioned Brick-walk in Baker's Hill	GK (Mar./Apr.) GK (Apr.)
1753	Zigzag constructed up hanger	Acc. (Sept. 1752/Oct. 1753)
1754	Plants box trees in vista Six gates in perspective	GK (Nov.) JM 90
1755	Attempts laurel and ilex hedge, and laurel screening of 'necessary'	GK (Mar./Apr.)
1756	Screens off butcher's yard with limes Opening of vista; and erection of oil-jars (incl. mounts & pedestals c. 9 ft high)	GK (Mar.) Acc. (Feb.); GK (May)
1757	Hercules board-statue cut	GK (Oct.)
1758	Sets up Hercules '20 Yards into the Hanger' Hermitage adjacent top of Zigzag, obelisk, and makes 'area'	GK (Jan.) Acc. (Apr./Oct.)

1759	Ha-ha (with bastion) made on Baker's Hill,
	   and 'fenced with sharp'ned piles'
	Constructs mount (5 ft high, 6 ft diameter
	   at top) 'at the bottom of ye great mead'
	   with 'wine-pipe' on top
		GK (Mar.)
		
		GK (Mar.); Acc.
		   (Mar./Aug.)

1760	Buys part of Lassam's orchard from John Wells	GK (June); Acc. (July)

Appendix C

<table>
<tr><td>YEAR</td><td>ITEM</td><td>REFERENCE</td></tr>
<tr><td>1761</td><td>Ha-ha 'of blue rags' in new garden</td><td>GK (Jan.)</td></tr>
<tr><td></td><td>Brick-walk to alcove (see 1762); lays large stone drain</td><td>GK (Mar.)</td></tr>
<tr><td></td><td>Terrace levelled and turfed</td><td>GK (Mar./Apr.)</td></tr>
<tr><td></td><td>Low circular mount (turfed) round great oak</td><td>GK (Apr.)</td></tr>
<tr><td></td><td>Forest-chair on bastion</td><td>GK (May)</td></tr>
<tr><td></td><td>Fruit-wall completed ('Total expence' — £38 10s. 1d.)</td><td>GK (July); Acc. (Nov.)</td></tr>
<tr><td></td><td>Sundial erected ('large Portland dial-post, & slab from Andover' and 'brass dial plate')</td><td>Acc. (Nov.)</td></tr>
<tr><td>1762</td><td>Alcove at end of terrace, with straw doors</td><td>JM 170; GK (Nov.)</td></tr>
<tr><td>1763</td><td>Speculation about constructing lake</td><td>JM 176</td></tr>
<tr><td>1764</td><td>Paints alcove</td><td>GK (May)</td></tr>
<tr><td></td><td>Fixes name-stone on fruit-wall</td><td>GK (Sept.)</td></tr>
<tr><td>1765</td><td>Plants fanned elms and laurels for screening</td><td>GK (Mar.)</td></tr>
<tr><td></td><td>Stable renovated</td><td>GK (June)</td></tr>
<tr><td>1766</td><td>Moves barn (structure 40 ft long)</td><td>GK (Apr.)</td></tr>
<tr><td>1773</td><td>Levels and turfs walks</td><td>NJ (Nov.)</td></tr>
<tr><td>1776</td><td>New Hermitage ('a plain cot') built</td><td>Bell, **2**: 126</td></tr>
<tr><td>1777</td><td>Buys from John Wells a 'little farm' — 'the fields behind my house, that *angulus iste* which the family have so long desired'</td><td>Bell, **2**: 62</td></tr>
<tr><td>1780</td><td>Bostal ('a sloping path') dug, 414 yds long</td><td>NJ (Sept.)</td></tr>
<tr><td>1783</td><td>Alcove built on Bostal</td><td>NJ (Oct.)</td></tr>
<tr><td>1788</td><td>Nails Greek and Italian inscriptions to front of alcove on hanger</td><td>NJ (Oct.)</td></tr>
<tr><td>1791</td><td>Erects large cross on Hermitage</td><td>NJ (Nov.)</td></tr>
<tr><td></td><td>New thatch on Hermitage</td><td>JR 1306/33</td></tr>
</table>

KEY: Acc. = Accounts at Gilbert White Museum, Selborne; Bell = Bell (1877); GK = *Kalendar*; JM = Mulso (1907); NJ = *Journal*; JR = John Rylands Library, Manchester University.

APPENDIX D

The Invitation to Selborne

White's poem, 'The Invitation to Selborne', was not merely a literary exercise: it was written as a real invitation and sent to friends to encourage them to visit Selborne. Of his several poems and translations it is his most sustained poetic production and, alongside Grimm's drawings, conveys vividly the topography of the village.

Reference to the poem is made in Chapter 2 (Note 33), Chapter 4 (Note 20) and Chapter 6 (Note 29). The text below is taken from *Selborne* (1813), an edition known as Mitford's edition because it included notes by the Reverend John Mitford of Benhall, Suffolk.

THE

INVITATION TO SELBORNE

See Selborne spreads her boldest beauties round
The varied valley, and the mountain ground,
Wildly majestic! what is all the pride
Of flats, with loads of ornament supply'd?
Unpleasing, tasteless, impotent expense,
Compar'd with nature's rude magnificence.
*   Arise, my stranger, to these wild scenes haste;*
The unfinish'd farm awaits your forming taste:
Plan the pavilion, airy, light and true;
Thro' the high arch call in the length'ning view;
Expand the forest sloping up the hill;
Swell to a lake the scant, penurious rill;
Extend the vista, raise the castle mound
In antique taste, with turrets ivy-crown'd;
O'er the gay lawn and flow'ry shrub dispread,
Or with the blending garden mix the mead;
Bid China's pale, fantastic fence, delight,
Or with the mimic statue-trap the sight.
*   Oft on some evening, sunny, soft and still,*
The Muse shall lead thee to the beech-grown hill,

To spend in tea the cool, refreshing hour,
Where nods in air the pensile, nest-like bower;
Or where the Hermit hangs the straw-clad cell,
Emerging gently from the leafy dell;
By fancy plann'd; as once th' inventive maid
Met the hoar sage amid the secret shade;
Romantic spot! from whence in prospect lies
Whate'er of landscape charms our feasting eyes;
The pointed spire, the hall, the pasture-plain,
The russet fallow, or the golden grain,
The breezy lake that sheds a gleaming light,
'Till all the fading picture fail the sight.

Each to his task; all different ways retire;
Cull the dry stick; call forth the seeds of fire;
Deep fix the kettle's props, a forky row,
Or give with fanning hat the breeze to blow.

Whence is this taste, the furnish'd hall forgot,
To feast in gardens, or th' unhandy grot?
Or novelty with some new charms surprizes,
Or from our very shifts some joy arises.
Hark, while below the village-bells ring round,
Echo, sweet nymph, returns the soften'd sound;
But if gusts rise, the rushing forests roar,
Like the tide tumbling on the pebbly shore.

Adown the vale, in lone, sequester'd nook,
Where skirting woods imbrown the dimpling brook.
The ruin'd Convent lies; here wont to dwell
The lazy canon midst his cloister'd cell;
While papal darkness brooded o'er the land,
Ere reformation made her glorious stand:
Still oft at eve belated shepherd-swains
See the cowl'd spectre skim the folded plains.

To the high temple would my stranger go,
The mountain-brow commands the woods below;
In Jewry first this order found a name,
When madding Croisades set the world in flame;
When western climes, urg'd on by Pope and priest,
Pour'd forth their millions o'er the deluged east;
Luxurious knights, ill suited to defy
To mortal fight Turcéstan chivalry.

Nor be the Parsonage by the muse forgot:
The partial bard admires his native spot;

Smit with its beauties, loved, as yet a child,
(Unconscious why) its scapes grotesque, and wild.
High on a mound th' exalted gardens stand,
Beneath, deep vallies scoop'd by nature's hand.
A Cobham here, exulting in his art,
Might blend the General's with the Gardener's part;
Might fortify with all the martial trade
Of rampart, bastion, fosse, and palisade;
Might plant the mortar with wide threat'ning bore,
Or bid the mimic cannon seem to roar.

   Now climb the steep, drop now your eye below,
Where round the blooming village orchards grow;
There, like a picture, lies my lowly seat,
A rural, shelter'd, unobserv'd retreat.

   Me far above the rest Selbornian scenes,
The pendent forests, and the mountain-greens
Strike with delight; there spreads the distant view,
That gradual fades till sunk in misty blue:
Here nature hangs her slopy woods to sight,
Rills purl between and dart a quivering light.

Blue Mist

The *Journal* records a 'Blue' mist on several occasions, one of which is 23 Feb. 1770:

> Blue mist. Vulg: [commonly] called London smoke . . . When such mists appear they are usually followed by dry weather. They have somewhat the smell of coal-smoke, & therefore are supposed to come from London, as they always come to us with a N:E wind.

White was right: 'smog' and 'peasoupers' associated particularly with anti-cyclonic conditions plagued London and other urban areas from Elizabethan times until the Clean Air Acts (1956, 1968); by the later decades of the eighteenth century fogs were so common (Brazell, 1968) that a Continental visitor contended that they 'must necessarily have an effect upon the constitution of the inhabitants' and attributed to the fogs a main cause of the 'English Melancholy . . . which so predominates in their constitution':

> The smoke of sea-coal fires, with which the atmosphere of London is generally filled, may be reckoned amongst the physical causes of the melancholy of its inhabitants. The terrestrial and mineral particles, with which that smoke is impregnated, insinuate themselves into the blood of those who are always inhaling them, render it dull and heavy, and carry with them new principles of melancholy (Grosley, 1772: 165, 166–7).

As for a cure, the most imaginative idea for alleviating the acrid odours and other effects of the sea-coal, which were so prevalent until the establishment of deep mining, had been made more than 100 years earlier by John Evelyn in a little-known tract entitled *Fumifugium: or, The Inconvenience of the Aer, and Smoake of London Dissipated* (1661). The substance of the proposal was to surround the city with plantations of 'such *Shrubs*, as yield the most fragrant and odoriferous *Flowers*, and are aptest to tinge the *Aer* upon every gentle emission at a great distance', an idea brought to fruition in the municipal parks of the nineteenth century and in the garden city and country park of the present century.

The plantings recommended by Evelyn, which White must surely have appreciated (brother Benjamin was to issue an edition of the tract in 1772), were as follows:

> That all low-grounds circumjacent to the City, especially *East* and *South-west*, be cast and contriv'd into square plots, or Fields of twenty, thirty, and forty *Akers*, or more, separated from each others by Fences of

double *Palisads*, or *Contr'spaliers*, which should enclose a Plantation of an hundred and fifty, or more, feet deep, about each Field; not much unlike to what His *Majesty* has already begun by the wall from old *Spring-garden* to *St. James's* in that *Park*; and is somewhat resembled in the new *Spring-garden* at *Lambeth*. That these *Palisads* be elegantly planted, diligently kept and supply'd, with such *Shrubs*, as yield the most fragrant and odoriferous *Flowers*, and are aptest to tinge the *Aer* upon every gentle emission at a great distance: Such as are (for instance amongst many others) the *Sweet-brier*, all the *Periclymenas* and *Wood-binds*; the Common *white* and *yellow Jessamine*, both the *Syringas* or *Pipe trees*; the *Guelder-rose*, the *Musk*, and all other *Roses; Genista Hispanica*: To these may be added the *Rubus odoratus, Bayes, Juniper, Lignum-vitae, Lavendar:* but above all, *Rosemary*, the *Flowers* whereof are credibly reported to give their scent above thirty Leagues off at Sea, upon the coasts of *Spain*; and at some distance towards the Meadow side, *Vines*, yea, *Hops*.

> —Et Arbuta passim,
> Et glaucas Salices, Casiamque Crocumque rubentem,
> Et pinguem Tiliam & ferrugineos Hyacinthos, &c. [Virgil]

For, there is a very sweet smelling *Sally*, and the blossoms of the *Tilia*, or *Lime-tree*, are incomparably fragrant; in brief, whatsoever is odoriferous and refreshing.

That the *Spaces*, or *Area* between these *Palisads*, and Fences, be employ'd in Beds and Bordune of *Pinks, Carnations, Cloves, Stock-gilly-flower, Primroses, Auriculas, Violets,* not forgetting the *White*, which are in flower twice a year, *April* and *August; Cowslips, Lillies, Narcissus, Strawberries,* whose very leaves as well as fruit, emit a *Cardiaque*, and most refreshing *Halitus*: also *Parietaria Lutea, Musk, Lemmon,* and *Mastick Thyme: Spike, Cammomile, Balm, Mint, Marjoram, Pempernel,* and *Serpillum,* &c. which upon the least pressure and cutting, breathe out and betray their ravishing odors.

That the Fields, and Crofts within these Closures, or invironing Gardens, be, some of them, planted with *wild Thyme*, and others reserved for Plots of *Beans, Pease* (not *Cabbages*, whose rotten and perishing stalks have a very noisom and unhealthy smell, and therefore by *Hippocrates* utterly condemned near great Cities) but such blossom-bearing Grain as send forth their virtue at farthest distance, and are all of them *marketable* at *London*; by which means, the *Aer* and *Winds* perpetually fann'd from so many circling and encompassing Hedges, fragrant Shrubs, Trees, and Flowers (the amputation and prunings of whose superfluities, may in *Winter*, on some occasions of weather, and winds, be burnt, to visit the City with a more benign *smoake*) not onely all that did approach the *Region*, which is properly design'd to be Flowery; but even the whole City, would be sensible of the sweet and ravishing varieties of the perfumes, as well as of the most delightful and pleasant objects, and places of Recreation for the Inhabitants (*Fumifugium*, ed. Joan Evans, n.d.; Swan Press; for TP, 1680).

APPENDIX F

Pennant's Queries

The systematic collection of information about natural history by means of circulating likely informants with a list of queries had been promoted by Robert Plot in the previous century (Gunther, 1939). Thomas Pennant adopted the practice prior to his tour of Scotland by arranging for his visit to be preceded by the following letter, the text of which is taken from The Scots Magazine, **34**, *April 1772. Record of White's attitude to the worth of the procedure — inevitably dependent upon the reliability of the informants — is not fully explicit, but his insistence in advice to his brother John on the importance of studying a territory first hand (see Chapter 11, Note 5) and his admiration for the work of Scopoli suggest substantive reservation amounting to a profound scepticism.*

To every Gentleman desirous to promote the Publication of an Accurate Account of the Antiquities, Present State, and Natural History of SCOTLAND.

SIR,

The great civility and hospitality I experienced in my journey through part of North Britain in 1769, encourage me to make a visit to the places I have not yet seen. Permit me to prepare you for my coming, by sending this notice of my intention of being in your neighbourhood the ensuing summer, and of paying my respects to you. As my stay can be but very short, I am desirous of being at once directed to the objects most worthy the observation of a traveller, and to be favoured with a collection of such things as I take the liberty of enumerating, and which it is impossible for a transient visitor to get together. As my sole objects are, my own improvement, and the true knowledge of your country, hitherto misrepresented, I have no doubt of your complying with my wishes, which are included in the following queries and requests.

1. What lakes, rivers, water-falls, or remarkable springs or fountains are there near you?

2. What remarkable natural caves, rocks, or mountains: what picturesque scenes worth drawing?

3. Any remarkable echoes? Have any remarkable storms, whirlwinds, earthquakes, or other singular phaenomena, happened in your neighbourhood?

4. What are the predominant diseases of your country? and what are the

means of cure used by your natural physicians? Have you lost any ancient distempers, or received any new ones?

5. Are there any people remarkable for their longevity, or for any particular accidents in body or mind? any bards? any ingenious self-taught artificers?

6. Does the number of the inhabitants in your parish increase or decrease? and to what cause may either be owing?

7. What may the present number be?

8. What are the principal commercial productions of your neighbourhood, manufactures, fisheries, sorts of corn raised? do you raise enough to export or otherwise?

9. What manures do you use?

10. Do you cultivate flax, hemp, potatoes, &c.?

11. What number of cattle, horses, sheep, or swine, are sent annually out of your parish?

12. What remarkable antiquities, artificial caves, mounts, intrenchments, druidical circles, pillars, or stones, crosses, grave-stones, monuments, inscriptions?

13. Any ruins of religious houses, or churches; or of ancient castles or cities? Favour me as much as possible with their history.

14. Please to inform me, Have any gentlemen in your neighbourhood any remarkable antiquities relating to the Scottish history in their possession; ancient weapons, stone, or iron; adder-stones, or glass-beads, brotches, or the like?

15. Has any gentleman any remarkable portraits, or other pictures?

16. Have any remarkable battles been fought in your parish?

17. Do the people retain any remarkable customs or superstitions?

18. What are their sports and plays? have they any annual festivals or commemorations?

19. Please to observe and collect against my coming, specimens of all sorts of earths, clays, stones, marbles, pebbles, sands, asbestus, minerals, and ores; also bituminous fossils, petrified shells, wood, fossil-wood, in short, fossils of all denominations.

20. Collect all sorts of fresh sea-shells, crabs, lobsters, corals, sea-plants, and every thing flung ashore by the waves.

21. Please to point out all singular trees and vegetables, and their uses.

22. Remark all the principal places where the sea-fowl resort to; also, where eagles and falcons breed. Please to preserve any singular birds by slaying, and stuffing them; and collect insects of all kinds.

The object of my collecting is not selfish; whatsoever I may acquire will be at the service of the national repository, the Musaeum at Edinburgh, now under the care of my friend Dr Ramsay, Professor of Natural History.

I flatter myself that you will accept this with candour, and that you will receive the writer as one who wishes well to your country in particular, and to mankind in general.

THOMAS PENNANT.
Downing, Flintshire, Jan. 31. 1772.

P.S. As it is impossible to send letters to every person of curiosity in the country, I beg you will take the trouble of communicating this to any one of your acquaintance, who, you think, can assist me in my inquiries; and it, in consequence of this letter, you prepare any remarks, or make any collection of the natural productions of your country, which you have not an opportunity to communicate to me while on my tour, be pleased to send them to Dr Ramsay, by whose means I can afterwards have the use of them.

APPENDIX G

Four Examples of White's Procedures in the Preparation of *Selborne*

Documentation has been provided earlier to show that White gave meticulous attention to the detail of his records, and it is somewhat puzzling to find in *Selborne* examples of what might be termed local inaccuracy. Four, indicative of some of his practices, will be commented on here.

(i) The first, a misreading of Barrington's handwriting, occurs in the 1770 October letter (Barrington 7) that he sent his correspondent from Ringmer. It begins:

> I am glad to hear that *Kuckalm* is to furnish you with the birds of *Jamaica*; a sight of the *hirundines* of that hot and distant island would be a great entertainment to me.

There is no later mention of these birds, so whether White ever saw them is unrecorded. It is known, however, that the provider was to be not an undocumented naturalist (Kuckalm) but T. S. Kuckhan, who presented the Royal Society in 1772 with 'Specimens of Flies and other Insects from Jamaica'.[1] In itself, an error of this kind may not be important, but it illustrates the isolation White suffered, both in his researches (he had few opportunities to *talk*, critically, with others) and in the revision of his manuscript when the time came to submit it to the printer. In part, this reflects the particular familial status of his publisher—his work was effectively seen through the press by his eager and willing, yet inexperienced, niece Mary and her husband—but it also points to the range of *Selborne* and to the absence of friends who were knowledgeable enough about the London scene and about the many aspects of natural history engrossed in White's work to have been able to advise him at the final stages of writing.

(ii) If an informed and attentive friend might have noticed the kind of slip given in (i) above, White himself or his relatives should fairly have picked up another. One of the late letters in the Barrington sequence (Barrington 49 dated May 1779) reports the shooting locally of five specimens of *Himantopus himantopus* (black-winged stilt); evidently, White was intrigued by the length and apparent fragility of the bird's legs and pursued in the letter the relationship between body weight and leg length. To emphasise his point he compared his relationship for the stilt with the same relationship in a flamingo (a bird he was familiar with from one of the Gibraltar cargoes),[2] and worked out that if the same proportions applied the flamingo would have legs over ten feet long! Unfortunately, the numerical calculation used to arrive at this answer was flawed. Marvelling at the stilt's legs, as if they were 'in *caricatura*' and designed at the fancy of a Chinese or Japanese draughtsman, he applied the linear calculation necessary for the measurement of the legs to the quite different cubic calculations necessary to work out the proportionate weights of the bird and concluded that, if the leg:weight ratio of the stilt obtained in a bird the size of a flamingo, then the result would be 'such a monstrous proportion as the world never saw!' Interestingly, instead of further speculation about such a bird, White continued his letter by remarking on the presence of the stilt in Britain, and noticed that there was no available explanation for its behaviour, whether accidental or motivated he is unsure. This is important, and of far greater significance for understanding the edge of White's thinking than the mathematical error—mischief of that kind can creep into the most attentive mind. When presented with fauna that were unfamiliar, it was customary for him to turn to measurement, the account of the moose at Goodwood (Pennant 28) being the most famous; but presented with behaviour that seemed inexplicable, it was equally customary, particularly in extreme instances, for him to appeal to Providence.[3] Notable examples of this last are the tortoise, whose capacity

> to squander more than two thirds of it's existence in a joyless stupor, and be lost to all sensation for months together in the profoundest of slumbers

he found a matter of great puzzlement; and the cuckoo, whose selective deposition of eggs he saw as illustrating

> in a fresh manner, that the methods of Providence [were] not subjected to any mode or rule, but astonish us in new lights, and in various and changeable appearances.[4]

Observation of the stilt in the hand and the consequent calculations would seem to justify a like appeal to Providence, but on this occasion it is not made—and its absence is mildly provocative! Why should an appeal be made to explain something about the tortoise, say, or the cuckoo, but not about the moose or the stilt? One answer would be to argue that such inconsistency of explanation was a matter of style: when faced with the inexplicable, persistent repetition of an appeal to Providence might appear jejune, even clumsy. Support for this view could be gathered from nearly every page of *Selborne* for, as has already been noted, it is the book's strongly varied presentational surface that contributes markedly to its appeal.[5] But there is an additional answer. This centres on White's engagement with reality: the longer he studied Nature in the flesh, the more closely he observed, the more convinced he became that any inexplicability was apparent only; in due time careful, accurate observation and experiment would yield results. He must have been convinced that, in the case of the stilt, if he had only been able to see a live bird some explanation of its ability to walk, in spite of its remarkable leg:weight proportions, would have been possible; and if data of the right kind had been available then the bird's very presence in Britain could also be explained.

Discoveries and solutions of this kind would, of course, be greatly promoted by accurate *Journal* data; and it is the relation of observations made in the *Journal* to the use of the very same data in *Selborne* that provides the remaining two examples.

(iii) It is clear that White frequently wrote with his *Journal* open before him. This is evident from careful comparison of many topics and even dates of observed phenomena in *Selborne* with entries made in the *Journal*. This is true both for the writing of real letters to Pennant and Barrington that underlie some, but not all, such letters in *Selborne* and for letters similarly addressed but prepared solely for publication. For example, in a letter to Pennant of 22 February 1770 (Pennant 27) there is mention of procuring a litter of hedgehogs in 'June last', and inspection of the *Journal* shows the litter was indeed obtained on 10 June 1769; further, in the sequence of letters about the climate at Selborne (Barrington 61–66) much of the detail, even to the height of the barometer to tenths of an inch, can be similarly confirmed. Indeed, in Barrington 62, which gives an account of 'the remarkable frost in *January* 1776', the author is so concerned to ensure the precision of his report that he informs his reader that the

> most certain way to be exact will be to copy the passages from my journal, which were taken from time to time as things occurred.

With this level of concern for precision of report again demonstrated, it is puzzling to discover several instances of inconsistency between the data given in the *Journal* and the same data (supposedly) mentioned in *Selborne*. The two examples I now give derive from close avian observation; a third—a most substantive one—has been treated in the main text and concerns a mammal. All three illuminate aspects of the relationship of the immediate daily record (notably the *Journal*) to the fashioned record (*Selborne*) and illustrate the caution that needs to be brought to a reading of the finished work.

The examples relating to birds occur in letters to Barrington (Barrington 2 and 7). The first of these letters, dated 2 November 1769, is based on a communication devoted primarily to listings of different kinds of singing birds that White sent his new correspondent on 9 November 1769. One of the lists is headed 'Birds that sing for a short time, and very early in the spring', and includes the mistle thrush, *Turdus viscivorus*. Against the entry a note is provided; in the real letter this reads as

> sings only in the early months of spring: is called in Hants the storm-cock: is the largest singing bird we have.[6]

In *Selborne* the note reads:

> January the 2d, 1770, in *February*. Is called in *Hampshire* and *Sussex* the storm-cock, because it's song is supposed to forebode windy wet weather: is the largest singing bird we have.

The inclusion of a precise date in this second note is unremarkable;[7] the real interest lies elsewhere— in the date provided being subsequent to the date of the letter. The accuracy of the observation is not in question: it is validated by the entry in the *Journal* for the same day. What is bemusing is that a letter of a certain date should include material that is later than that date. What seems to have happened is that at some stage of revising his real letters for publication White decided to add to the letter, in the interest of greater precision for such a notable occurrence, an indisputable reference. To modern eyes the provision of data of this kind may seem mistaken, and open to dispute: earlier or, as the case may be, later dates must always be possible. But, and this is another awareness in an understanding of the relation of the *Journal* to *Selborne*, we need to recall the role adopted by White at the time he first became acquainted with Barrington. In the preface to the blank *Journal* which he had received from its designer in 1767, Barrington had written:

> from many such journals kept in different parts of the kingdom, perhaps the very best and most accurate materials for a General Natural History of Great Britain may in time be expected . . .

In beginning his correspondence with Barrington in 1769, and in using the *Journal* as a source, White must be seen as contributing to this 'General Natural History'. This is a role White kept before himself and one we also need to be aware of when reading him. The data he provided could be refined through comparison with data furnished by other partial observers; and it is on those terms that the dates given by White must be seen, the phenomena themselves being observed, for the most part, solely within his native parish.

(iv) If, however, White's main observational context was the parish of Selborne, his writing was not always conducted within the parish. Some of the letters that underlie *Selborne* were written at Fyfield during visits to brother Henry, some in London, and some before 1780 while he stayed with Mrs Snooke at Ringmer; it is one of these last that provides the second instance of inconsistency between White's two kinds of record. Writing to Barrington on 8 Oct. 1770, White remarked:

> In July I saw several cuckows skiming over a large pond; & found after some observation that they were feeding on the libellulae or dragon-flies; some of which they caught as they settled on the weeds, & some as they were on the wing. Notwithstanding what Linnaeus says, I can't be induced to believe that they are birds of prey.[8]

In fact, the remarks were written from memory. The *Journal* for July 1770 mentions the cuckoo three times: entries on 1 July (at Fyfield) and 7 July (at Selborne) report hearing the bird sing, and a long note on 25 July relates seeing 'a young cuckow . . . in the nest of a tit lark [*Anthus pratensis*]'. In the 8 October letter, it is a version of this last note, without mention of a month, that immediately precedes White's remarks on seeing cuckoos feeding 'over a large pond': evidently, the month of observation had been recalled too late, and became attached to the wrong event. Confirmation that this is so is given in the *Journal*, for at the beginning of September 1770 is a footnote entry:

> Sept: 1st. Cuckows skim over the ponds at Oakhanger [two miles east of Selborne], & catch libellulae on the weeds, & as they flie in the air. I can give no credit to the notion that they are birds of prey. They have a weak bill & no talons.

Inaccuracy of this kind, it might be thought, would have been noticed by White at the time he came to prepare his letters for publication, checking the accuracy of all dates in a work purporting to be history as well as science being a necessary element in any editing process. But, read alongside the most blatant exemplary instance of inconsistency between the records, that of the otter, it must be agreed that the editing of material for *Selborne* is also responsible, perhaps deliberately so, for it being possible to claim that the two kinds of record need to be read differently.

NOTES

1. Noted in the *Philosophical Transactions*, vol. 62 (1772), against the date 6 Feb. 1772; a similar gift was noted for 28 Apr. 1774, and earlier (1770) the Society had published a paper of his on the preservation of dead avian specimens in natural attitudes.

2. And so delighted by it that a specimen was presented to the Royal Society: see Chap. 10, text associated with Notes 25 and 26.

3. White's mentors in respect of Providence are found in the works of Ray and Derham particularly, but the tradition continued well into the time of White's maturity, and a poetic treatment of 'the superintendency of Providence' occurs in Ogilvie (1764).

4. For the tortoise, see *Selborne*, Barrington 50 (dated 21 Apr. 1780), and for the cuckoo Barrington 4 (dated 19 Feb. 1770).

5. A sequential treatment of this aspect of *Selborne* is given in Keith (1975).

6. B.L. Add. MS 31852, White to Barrington, 9 Nov. 1769.

7. A similar precision, although without mention of a year, occurs in other lists: for example, the yellowhammer is recorded as singing until 21 Aug., and the goldfinch until 16 Sept.

8. Selborne MS, White autograph.

Bibliography

Use of both manuscript and printed sources leads to some variability in presentation. Where an accurate printed text is available it has been consulted and referenced accordingly. Some texts, however, were prepared fifty or more years ago, at a time when the expectations attaching to transcription were different from those obtaining today; in such instances I have gone direct to manuscript sources and observed the following principles. Orthography, except for an occasional editorial clarification or comment in square brackets, is that given in the original manuscripts: for example, ampersands, 'ye' (for 'the') and 'it's' (for 'its') are retained; punctuation, including the eighteenth-century preference for a colon where modern usage would require a semi-colon or full period, and paragraphing are unchanged; deletions are unrecorded. In quoting from the *Journal* most entries can be ascribed to a particular date, but in a few instances, most commonly entries that occur at the foot of the page, this is not possible and such passages are given as 'at' (rather than 'on') such and such a date.

The listing of books and articles includes everything that has materially contributed to the present study. Generally, dates given are for first editions; where this is not the case the date of the first edition is in square brackets; publishers are listed only for books issued in the past thirty years. I much regret that the second volume of the Century Hutchinson edition of White's journals has, at the time of writing, not been available for consultation: it should prove a valuable publication.

MANUSCRIPT SOURCES

White autographs

Gilbert White Museum, Selborne — National Register of Archives 11555
    Account books
        These include details of expenses in 1740s and 1750s; housekeeping 'from the Death of my Father'; rents; charities; and transactions with Benjamin White (brother) in London
    Brewing records
    Selborne MS (the manuscript from which the first edition was set by the printer)
    Sermons

British Library
  Garden-Kalendar B.L. Add. MS 35139
  Naturalist's Journal B.L. Add. MS 31846–31851
  Letters to Thomas Pennant B.L. Add. MS 35138
  Letters to Daines Barrington B.L. Add. MS 31852
Selborne Society, London
  Calendar of Flora (Flora Selborniensis)
Henshaw Collection, Houghton Library, Harvard University—bMS Eng 731
  Letters to Thomas Pennant
  Letters to John White (brother)
John Rylands Library, Manchester University—Eng MSS 1306
  Letters to members of the family
Fitzwilliam Museum, Cambridge
  Letters to Mary White (niece)
Oriel College, Oxford—Case D C III. 11–13
  Account books
    These include information on the 'Executorship at Thorney' July 1746; loans to members of the family; wine bill for 1779
  Letter to Mrs John White
  Letter to Thomas Barker
Royal Society, London—Letters and Papers VI.27, 93
  Essays on the hirundines (not in holograph)
Record Offices, Hampshire and Wiltshire
  Parish registers, including those for Selborne, Faringdon, and West Dene

Other autographs

Gilbert White Museum, Selborne
  Legal memoranda of John White (father)
  Commonplace book of Thomas White (brother)
  Letters from Mary White (niece) to her brother
British Library
  Journals of Henry White (brother)
Selborne Society, London
  Letter from James Gibson
  Letters from William Sheffield
John Rylands Library, Manchester University
  Letters from John Mulso
National Library of Wales, Aberystwyth
  Thomas Pennant papers
Alexander Turnbull Library, New Zealand
  Letters from Joseph Banks to Thomas Pennant
Corpus Christi College, Oxford
  Papers concerning expulsion of John White (brother)
Lancing College, West Sussex
  Meteorological diaries of Thomas Barker

Bibliography

Leicestershire Record Office
    Barker papers
National Library of Scotland, Berkshire County Record Office, and Suffolk Record Office
    Daines Barrington papers

BOOKS AND ARTICLES

ABERCROMBIE, John. 1769 [1767]. *Every Man his own Gardener*. London.

ADANSON, M. 1759. *A Voyage to Senegal, the Isle of Goree, and the River Gambia*. London.

ADDISON, William. 1947. *The English Country Parson*. J. M. Dent, London.

ALBIN, Eleazar. 1737. *A Natural History of English Song-birds*. London.

ALLEN, David Elliston. 1976. *The Naturalist in Britain: a Social History*. Allen Lane, London.

ANDERSON, Howard *et al.* 1968. *The Familiar Letter in the Eighteenth Century*. University Press of Kansas, Lawrence.

ASPÖCK, H. *et al.* 1980. *Die Neuropteren Europas*, 2 vols. Goecke & Evers, Krefeld.

ATTENBOROUGH, David. 1987. 'How unnatural is TV natural history?', Huw Wheldon Memorial Lecture, extracts in *The Listener* 7 May 1987.

BALDWIN, Stuart A. 1986. *John Ray (1627–1705) Essex Naturalist: a Summary of his Life, Work and Scientific Significance*. Baldwin's Books, Witham.

BARKER, Thomas. 1749. 'An account of an extraordinary meteor', *Phil. Trans. R. Soc. London* **46**: 248–9.

—— 1757. *An Account of the Discoveries concerning Comets*. London.

—— 1771. 'Meteorological observations at Lyndon, Rutland', *Phil. Trans. R. Soc. London* **61**: 221–6, 227–9.

BARLOW, Edward. 1722 [1717]. *An Exact Survey of the Tide*. London.

BARRETT, Alan. 1987. 'Sheffield, the unknown provost', *Worcester College Record*: 23–34.

[BARRINGTON, Daines] 1767. *The Naturalist's Journal*. London.

—— 1773. 'Experiments and observations on the singing of birds', *Phil. Trans. R. Soc. London* **63**: 249–91.

—— 1781. *Miscellanies*. London.

BATEY, Mavis. 1986 [1982]. *Oxford Gardens: the University's Influence on Garden History*. Scolar Press, London.

BEAGLEHOLE, J. C. (ed.) 1977 [1962]. *The 'Endeavour' Journal of Joseph Banks 1768–1771*, 2 vols. Trustees of the Public Library of New South Wales, in association with Angus and Robertson, Sydney.

BECKFORD, Peter. 1782 [1781]. *Thoughts upon Hunting*. Sarum.

BELL, Thomas (ed.) 1877. *The Natural History and Antiquities of Selborne*, 2 vols. London.

BERRY, R. J. 1987. 'Scientific natural history: a key base to ecology', *Biological Journal of the Linnean Society* **32**: 17–29.

BLENCOWE, Robert W. 1848. 'Extracts from the journal and account book of the Rev. Giles Moore', *Sussex Archaeological Collections* **1**: 65–127.

BRADLEY, Richard. 1719 [1717–18]. *New Improvements of Planting and Gardening.* London.

—— 1720 [1718]. *The Gentleman and Gardener's Kalendar.* London.

—— 1727. *A Complete Body of Husbandry.* London.

—— 1980 [1727–32]. *The Country Housewife and Lady's Director:* a facsimile of 1736 edn, ed. Caroline Davidson. Prospect Books, London.

BRAZELL, J. H. 1968. *London Weather.* HMSO, London.

BROWNE, Janet. 1983. *The Secular Ark: Studies in the History of Biogeography.* Yale Univ. Press, New Haven.

BURTON, Clive. 1979. *The Diaries of an 18th. Century Parson.* J. A. C. Sarsen, Andover.

CARR, D. J. (ed.) 1983. *Sydney Parkinson: Artist of Cook's 'Endeavour' Voyage.* Croom Helm, Beckenham.

CARTWRIGHT, James Joel (ed.) 1898–9. *The Travels through England of Dr. Richard Pococke . . . during 1750, 1751, and later Years*, 2 Vols. Camden Society, London.

CASH, R. C. 1963. 'A connection between Charles and Gilbert White', *Medical History* **7**: 273–5.

CHATFIELD, J. E. 1987. 'Likenesses of the Reverend Gilbert White', *Proceedings of the Hampshire Field Club and Archaeological Society* **43**: 207–17.

CHILDS, K. A. 1949. *Herbarium List.* Curtis Museum, Alton.

CLARIDGE, John. 1946 [1670]. *The Country Calendar or the Shepherd of Banbury's Rules*, ed. G. H. T. Kimble. Sylvan Press, London.

CLAY, Rotha Mary. 1941. *Samuel Hieronymus Grimm of Burgdorf in Switzerland.* London.

COBBETT, William. 1979 [1822]. *Cottage Economy.* Oxford Univ. Press, Oxford.

COMBRUNE, M. 1762. *Theory and Practice of Brewing.* London.

COOMBS, Franklin. 1978. *The Crows: a Study of the Corvids of Europe.* Batsford, London.

CORTÉS, J. E. *et al.* 1980. *The Birds of Gibraltar.* Gibraltar Bookshop, Gibraltar.

[CUNNINGHAM, Timothy] 1766 [1764]. *A New Treatise on the Laws for Preservation of the Game.* London.

CURTIS, W. Hugh. 1941. *William Curtis 1746–1799.* Warren, Winchester.

DAY, Robert Adams. 1966. *Told in Letters: Epistolary Fiction before Richardson.* University of Michigan Press, Ann Arbor.

DEFOE, Daniel. 1704. *The Storm.* London.

—— 1983 [1724–7]. *A Tour through Great Britain*, 3 vols. Folio Society, London.

DERHAM, W. 1714 [1713]. *Physico-Theology: or, a Demonstration of the Being and Attributes of God, from his Works of Creation.* London.

EDWARDS, George. 1972 [1770]. *Essays upon Natural History:* a facsimile. Paul P. B. Minet, Newport Pagnell.

EMDEN, Cecil S. 1948. *Oriel Papers.* Clarendon Press, Oxford.

—— 1956. *Gilbert White in his Village.* Oxford Univ. Press, Oxford.

EVELYN, John n.d. [1661]. *Fumifugium: or, the Inconvenience of the Aer, and Smoake of London Dissipated:* ed. Joan Evans. Swan Press, S.L.

—— 1801 [1664]. *Silva: or a Discourse of Forest-Trees*: ed. A. Hunter. York.

—— 1983 [1664–6]. *Kalendarium Hortense*: ed. Rosemary Verey. Stourton Press, Hackney, London.

FARBER, Paul Lawrence. 1982. *The Emergence of Ornithology as a Scientific Discipline 1760–1850*. D. Reidel, Dordrecht.

FLEMING, Laurence, and GORE, Alan. 1979. *The English Garden*. Michael Joseph, London.

FLETCHER, Dennis. 1976. '*Candide* and the theme of the happy husbandman', *Studies on Voltaire and the Eighteenth Century* **161**: 137–47.

FOSTER, J. 1888. *Alumni Oxonienses 1715–1886*, 4 vols. Oxford Univ. Press. Oxford.

FOSTER, Paul G. M. 1985a. 'The Gibraltar correspondence of Gilbert White', *Notes and Queries* **32**: 227–36, 315–28, 489–500.

—— 1985b. 'William Sheffield: four letters to Gilbert White', *Archives of Natural History* **12**: 1–21.

—— 1986a. 'The truffle records of Gilbert White', *Bulletin of the British Mycological Society* **20**: 42–7.

—— 1986b. 'The Hon. Daines Barrington, F.R.S.—Annotations on two journals compiled by Gilbert White', *Notes and Records of the Royal Society of London* **41**: 77–93.

—— 1986c. 'Quebec 1759: James Gibson, naval chaplain, writes to the naturalist, Gilbert White', *Journal of the Society for Army Historical Research* **64**: 218–23.

—— 1988. 'Dating of Gilbert White's botanical record', *Bulletin of the British Society for Eighteenth-Century Studies* No. 14.

FREEMAN, R. B. 1980. *British Natural History Books*. Dawson, Folkstone: Archon Books, Hamden, Connecticut.

FUSSELL, G. E. 1950. *More Old English Farming Books from Tull to the Board of Agriculture 1731 to 1793*. Crosby Lockwood, London.

GALPINE, John Kingston. 1983. *The Georgian Garden: an Eighteenth-Century Nurseryman's Catalogue*. Dovecote Press, Wimborne.

GARNER, W. W., and ALLARD, H. A. 1920. 'Effect of the relative lengths of day and night . . . on growth and reproduction of plants', *Journal of Agricultural Research* **18**: 553–606.

GIBSON, Donald (ed.) 1982. *A Parson in the Vale of the White Horse: George Woodward's Letters from East Hendred, 1753–1761*. Alan Sutton, Gloucester.

GILMOUR, J. S. L. (ed.) 1972. *Thomas Johnson: Botanical Journeys in Kent & Hampstead—a facsimile . . . with Introduction and Translation of his Iter Plantarum 1629 [&] Descriptio Itineris Plantarum 1632*. Hunt Botanical Library, Pittsburgh.

GLASS, Bentley, *et al*. (ed.) 1959. *Forerunners of Darwin: 1745–1859*. Johns Hopkins Univ. Press, Baltimore.

GOULD, Stephen Jay. 1980 [1978]. *Ever since Darwin: Reflections in Natural History*. Penguin, Harmondsworth.

GRAINGER, Margaret. 1983. *The Natural History Prose Writings of John Clare*. Clarendon Press, Oxford.

—— and WILLIAMSON, Richard. 1988. *The Nightjar*. Bishop Otter Trustee, Chichester (in press).

GREENOAK, Francesca (ed.) 1986. *The Journals of Gilbert White*, vol. 1 1751–1773. Century Hutchinson, London.

GRIFFIN, Dustin. 1986. *Regaining Paradise: Milton and the Eighteenth Century*. Cambridge Univ. Press, Cambridge.

GROSLEY, Pierre Jean. 1772. *A Tour to London*, 2 vols. London.

GUNTHER, R. T. 1939. *Dr. Plot and the Correspondence of the Philosophical Society of Oxford*, vol. 12 of *Early Science in Oxford*. Oxford.

HALES, Stephen. 1969 [1727]. *Vegetable Staticks*: ed. M. A. Hoskin. Macdonald, London: American Elsevier, New York.

—— 1731–3. *Statical Essays*, 2 vols. London.

—— 1758 [1743]. *A Treatise on Ventilators*. London.

HARVEY, John. 1972. *Early Gardening Catalogues*. Philimore, Chichester.

—— 1974. *Early Nurserymen*. Phillimore, Chichester.

—— 1981. *Early Horticultural Catalogues: Supplement to 1973 Checklist*. Univ. of Bath Library, Bath.

HELLER, John L. 1983. 'Notes on the titulature of Linnean dissertations', *Taxon* **32**: 218–52.

HENREY, Blanche. 1975. *British Botanical and Horticultural Literature before 1800*, 3 vols. Oxford Univ. Press, London.

HIGHET, Gilbert. 1957. *Poets in a Landscape*. Hamish Hamilton, London.

HILL, John. 1756. *The British Herbal*. London.

HITT, Thomas. 1755. *A Treatise of Fruit-Trees*. London.

HOARE, Michael E. 1976. *The Tactless Philosopher: Johann Reinhold Forster (1729–98)*. Hawthorn Press, Melbourne.

HOLT-WHITE, Rashleigh. 1901. *The Life and Letters of Gilbert White of Selborne*, 2 vols. London.

HOPKINS, Harry. 1985. *The Long Affray: the Poaching Wars 1760–1914*. Secker and Warburg, London.

HORACE. 1978 [1964]. *The Odes*: tr. James Michie. Penguin, Harmondsworth.

HORN, Pamela. 1981. *A Georgian Parson and Village: the Story of David Davies (1742–1819)*. Beacon, Abingdon.

HOWARD, Luke. 1803. 'On the modification of clouds', *Philosophical Magazine* **16**: 97–107.

HUDSON, William. 1762. *Flora Anglica*. London.

HUNT, John Dixon. 1976. *The Figure in the Landscape: Poetry, Painting, and Gardening during the Eighteenth Century*. Johns Hopkins Univ. Press, Baltimore.

—— 1986. *Garden and Grove: the Italian Renaissance in the English Imagination—1600–1750*. J. M. Dent, London.

HUNTER, A. 1777. *Georgical Essays*. York.

HUXHAM, John. 1739. *Observations de Aëra, et Morbis epidemicis*. Londini.

JOHNSON, Samuel. 1770 [1755]. *A Dictionary of the English Language*, 2 vols. London.

JOHNSON, Walter. 1928. *Gilbert White: Pioneer, Poet, and Stylist*. London.

KEITH, W. J. 1975. *The Rural Tradition*. Harvester, Hassocks.

KINGTON, J. A., and KINGTON, Beryl D. 1981. 'Thomas Barker of Lyndon Hall, Rutland and his weather observations', *Rutland Record* No. 2: 58–67.

KNIGHT, David M. 1972. *Natural Science Books in English 1600–1900*. Batsford, London.

LANGLEY, Batty. 1728. *New Principles of Gardening*. London.

LE FAYE, Deirdre. 1975. 'Selborne Priory, 1233–1486', *Proceedings of the Hampshire Field Club & Archaeological Society* **30**: 47–71.

LE ROUGETEL, Hazel. 1971. 'A Garden Kalendar 1751–1771: Gilbert White consults Philip Miller', *Journal of the Royal Horticultural Society* **96**: 412–18.

LIGHTFOOT, J. 1777. *Flora Scotica*, 2 vols. London.

LINNAEUS, Carl. 1956 [1735]. *Systema Naturae*, vol. 1: a facsimile of 10th edn. 1758. Trustees of British Museum (Natural History), London.

—— 1957 [1753]. *Species Plantarum:* a facsimile of the 1st edn, ed. W. T. Stearn, 2 vols. Ray Society, London.

LLOYD, Christopher. 1985 [1970]. *The Well-tempered Garden*. Viking, London.

LOCKLEY, Ronald. 1976 [1954]. *Gilbert White*. White Lion, London.

LOUDON, J. C. (ed.) 1872 [1829]. *An Encyclopaedia of Plants*. London.

LOVEJOY, A. O. 1933. *The Great Chain of Being*. Harvard Univ. Press, Cambridge, Mass.

LYON, John, and SLOAN, Phillip (eds.) 1981. *From Natural History to the History of Nature: Readings from Buffon and his Critics*. University of Notre Dame Press, London.

MABEY, Richard. 1986. *Gilbert White*. Century Hutchinson, London.

MACK, Maynard. 1969. *The Garden and the City*. University of Toronto Press, Toronto.

McKIE, Douglas. 1972. 'The scientific periodical from 1665 to 1798' and 'Scientific societies to the end of the eighteenth century', in *Natural Philosophy through the 18th Century and Allied Topics*, ed. Allan Ferguson. Taylor and Francis, London.

MALINS, Edward. 1966. *English Landscaping and Literature 1660–1840*. Oxford Univ. Press., Oxford.

MARKHAM, Sarah. 1984. *John Loveday of Caversham 1711–1789: the Life and Tours of an Eighteenth-Century Onlooker*. Michael Russell, Salisbury.

MARSHALL, John (tr.) 1911. *Horace's Complete Works*. J. M. Dent, London.

MARTIN, Edward A. 1934 [1897]. *A Bibliography of Gilbert White*. Folkestone.

MARTIN, Peter. 1984. *Pursuing Innocent Pleasures: the Gardening World of Alexander Pope*. Archon Books, Hamden, Connecticut.

MASON, William. 1819 [1772–9]. *The English Garden*. London.

MAYR, Ernst. 1982. *The Growth of Biological Thought: Diversity, Evolution, and Inheritance*. Harvard Univ. Press, Cambridge, Mass.

MEIRION-JONES, G. I. 1983. 'The Wakes, Selborne: an architectural study', *Proceedings of the Hampshire Field Club and Archaeological Society* **39**: 145–69.

MIDDLETON, W. E. Knowles. 1964. *The History of the Barometer*. Johns Hopkins Univ. Press, Baltimore.

—— 1966. *A History of the Thermometer*. Johns Hopkins Univ. Press, Baltimore.

MILLER, Philip. 1732 [1731]. *The Gardeners Kalendar*, 1st edn. London.

—— 1752 [1731]. *The Gardeners Dictionary*, 6th edn. London.

—— 1768 [1731]. *The Gardeners Dictionary*, 8th edn. London.

Milton, John. 1958. 'A Maske [*Comus*] Presented at Ludlow Castle' [1637] and 'Il Penseroso' [1645], in *The Poetical Works of John Milton*, ed. Helen Darbishire. Clarendon Press, Oxford.

Moir, E. A. L. 1964. *The Discovery of Britain: the English Tourists, 1540 to 1840*. Routledge and Kegan Paul, London.

Montagu, George. 1802. *Ornithological Dictionary*. London.

Morton, A. G. 1981. *History of Botanical Science*. Academy Press, London.

Mullet, Charles F. 1969. '*Multum in parvo*: Gilbert White of Selborne', *Journal of the History of Biology* **2**: 363–89.

Mulso, John. 1907. *The Letters to Gilbert White of Selborne*: ed. Rashleigh Holt-White. London.

Museum Rusticum et Commerciale. 1764, vol. 2. London.

Namier, Lewis, and Brooke, John. 1964. *The House of Commons 1754–1790*. HMSO for the History of Parliament Trust, London.

Nichols, John. 1966 [1812–15]. *Literary Anecdotes of the Eighteenth Century:* a facsimile, 9 vols. AMS Press, New York.

Noblett, William. 1982. 'Pennant and his publisher; Benjamin White, Thomas Pennant and *Of London*', *Archives of Natural History* **11**: 61–8.

Ogborn, Jon. 1985. *Voices of Science*. London Institute of Education, London.

Ogilby, John, and Morgan, William. 1770 [1759]. *The Traveller's Pocket-book*. London.

Ogilvie, John. 1764. *Providence: an Allegorical Poem*. London.

Parkinson, John. 1629. *Paradisi in Sole: Paradisus Terrestris:* a facsimile. Theatrum Orbis Terrarum, Amsterdam: W. J. Johnson, New York.

Pennant, Thomas. 1761–6. *British Zoology*. Cymmrodorion Society, London.

—— 1771. *Synopsis of Quadrupeds*. Chester.

—— 1771. *A Tour in Scotland 1769*. Chester.

—— 1771. 'Description of three curious fishes', *The Gentleman's Magazine* **41**: 249.

—— 1777. 'A Sketch of Caledonian Zoology', in J. Lightfoot, *Flora Scotica*.

—— 1948 [MS 1765–?]. *Tour on the Continent*, ed. G. R. de Beer. Ray Society, London.

Perkins, Walter Frank. 1961. *Catalogue of the Walter Frank Perkins Agricultural Library*. The University Library, Southampton.

Petti, Anthony G. 1977. *English Literary Hands from Chaucer to Dryden*. Edward Arnold, London.

P[lot], R[obert] 1677. *The Natural History of Oxford-shire*. Oxford and London.

Pope, Alexander. 1961 [1713]. 'Windsor-Forest', in *Pastoral Poetry and An Essay on Criticism*, ed. E. Audra and Aubrey Williams. Methuen, London: Yale Univ. Press, New Haven.

Porter, Roy. 1982. *English Society in the Eighteenth Century*. Penguin, Harmondsworth.

Prest, John. 1981. *The Garden of Eden: the Botanic Garden and The Re-Creation of Paradise*. Yale Univ. Press, New Haven.

Rannie, David Watson. 1900. *Oriel College* (University of Oxford College Histories). London.

RAVEN, Charles E. 1942. *John Ray Naturalist: his Life and Works*. Cambridge Univ. Press, Cambridge.

RAY, John. 1674. *A Collection of English Words*. London.

—— 1710. *Historia Insectorum*. Londini.

—— 1713. *Joannis Raii Synopsis Methodica Avium & Piscium*. Londini.

—— 1718. *Philosophical Letters*, ed. W. Derham. London.

—— 1750 [1691]. *The Wisdom of God Manifested in the Works of Creation*, 12th edn. Glasgow.

—— 1973 [1690]. *Synopsis Methodica Stirpium Britannicarum*: a facsimile of 3rd edn, 1724. Ray Society, London.

—— 1981 [1675]. *Dictionariolum Trilingue*: a facsimile with an introduction by W. J. Stearn. Ray Society, London.

RITTERBUSH, Philip C. 1964. *Overtures to Biology: the Speculations of Eighteenth-Century Naturalists*. Yale Univ. Press, New Haven.

ROLFE, W. D. Ian. 1983. 'William Hunter (1718–1783) on Irish "elk" and Stubbs's *Moose*', *Archives of Natural History* **11**: 263–90.

ROSE, Francis. 1972. 'Field Botanizing in 1629 . . .', in *Thomas Johnson: Botanical Journeys in Kent & Hampstead*, ed. J. S. L. Gilmour. Hunt Botanical Library, Pittsburgh.

ROUSSEAU, G. S. 1978. 'Science', in *The Eighteenth Century*, ed. Pat Rogers. Methuen, London.

—— 1978. 'Science books and their Readers in the Eighteenth century', in *Books and their Readers in Eighteenth-Century England*, ed. Isabel Rivers. Leicester Univ. Press, Leicester: St. Martin's Press, New York.

RYE, Anthony. 1970. *Gilbert White and his Selborne*. William Kimber, London.

SACHS, Julius. 1865. *Experimental-Physiologie*. Leipzig.

SALAMAN, Redcliffe N. 1949. *The History and Social Influence of the Potato*. Cambridge.

SAMBROOK, James. 1986. *The Eighteenth Century: the Intellectual and Cultural Context of English Literature 1700–1789*. Longmans, London.

SCOPOLI, Ioannis Antonii. 1769. *Annus I Historico-Naturalis*. Lipsiae.

SCOTT, Walter S. 1946. *White of Selborne and his Times*. John Westhouse, London.

SEARLE, S. A., and MACHIN, B. J. 1968 [1957]. *Chrysanthemums the Year Round*. Blandford Press, London.

SEDGWICK, Romney. 1970. *The House of Commons 1715–1754*. HMSO for the History of Parliament Trust, London.

SHENSTONE, William. 1778. *The Poetical Works of William Shenstone . . . with the Life of the Author and a Description of The Leasowes*, 2 vols. London.

SHERBO, Arthur. 1985. 'The English weather, *The Gentleman's Magazine*, and the brothers White', *Archives of Natural History* **12**: 23–9.

SIMONS, Arthur J. 1942. *The Vegetable Growers' Handbook*. Bakers of Codsall, Wolverhampton.

SLACK, A. A. 1986. 'Lightfoot and the exploration of the Scottish flora', *The Scottish Naturalist* (Botanical Society of the British Isles Conference Report, No. 20): 59–76.

SMITH, A. W. 1972. *A Gardener's Dictionary of Plant Names*. Cassell, London.

SMITH, Bernard. 1985 [1960]. *European Vision and the South Pacific*. Yale Univ. Press, New Haven.

SPENCE, Joseph. 1747. *Polymetis*. London.

STEARN, W. T. 1957. 'An Introduction to the *Species Plantarum* and cognate botanical works of Carl Linnaeus', in Linnaeus, *Species Plantarum* (1753): a facsimile of the 1st edn. Ray Society, London.

—— 1972. 'An Introduction to Vernacular Names', in A. W. Smith, *A Gardener's Dictionary of Plant Names*.

—— 1973. 'Ray, Dillenius, Linnaeus and the *Synopsis Methodica Stirpium Britannicarum*', in J. Ray, *Synopsis Methodica Stirpium Britannicarum* (1724): a facsimile of the 3rd edn. Ray Society, London.

STILLINGER, Jack. 1957. 'Gilbert White to Thomas Pennant: two original letters at Harvard', *Harvard Library Bulletin* **11**: 303–16.

STILLINGFLEET, Benjamin. 1762 [1759]. *Miscellaneous Tracts relating to Natural History, Husbandry, and Physick*. London.

SUTHERLAND, J. R. 1948. *A Preface to Eighteenth-Century Poetry*. Clarendon Press, Oxford.

SUTHERLAND, Lucy. 1973. *The University of Oxford in the Eighteenth Century: a Reconsideration*, a James Bryce Memorial Lecture. Oxford Univ. Press, Oxford.

SWAMMERDAM, Jan. 1758 [1737–8]. *The Book of Nature*. London.

[SWIFT, Jonathan] 1704. *A Tale of a Tub . . . to which is added, An Account of a Battle between the Ancient and Modern Books in St. James's Library*. London.

[——] 1726. *Travels into Several Remote Nations of the World by Lemuel Gulliver*. London.

TATE, Peter. 1986. *The Swallow*. Shire Natural History, Princes Risborough.

THACKER, Christopher. 1985 [1979]. *The History of Gardens*. Croom Helm, Beckenham.

—— 1983. *The Wildness Pleases: the Origins of Romanticism*. Croom Helm, Beckenham: St. Martin's Press, New York.

THOMAS, Keith. 1983. *Man and the Natural World: Changing Attitudes in England 1500–1800*. Allen Lane, London.

THOMSON, James. 1981 [1730]. *The Seasons*, ed. James Sambrook. Clarendon Press, Oxford.

TITTENSOR, A. M., and TITTENSOR, Ruth M. 1986. *The Rabbit Warren at West Dean Near Chichester*. Ruth and Andrew Tittensor Consultancy, Arundel.

TOWNSEND, F. 1904–29. *Flora of Hampshire*, London.

TROWBRIDGE, Hoyt. 1979. 'White of Selborne: the ethos of probabilism', in *Probability, Time, and Space in Eighteenth-Century Literature*, ed. Paula R. Backscheider. AMS Press, New York.

TURBERVILLE, A. S. (ed.) 1952 [1933]. *Johnson's England*, 2 vols. Oxford.

TURNER, Roger. 1985. *Capability Brown and the Eighteenth-Century English Landscape*. Weidenfeld and Nicolson, London.

TURNER, Thomas. 1984. *The Diary of Thomas Turner 1754–1765*, ed. David Vaisey, Oxford Univ. Press, Oxford.

VAUGHAN, John. 1906. *The Wild-Flowers of Selborne and other Papers*. John Lane, The Bodley Head, London.

VIRGIL. 1966. *The Eclogues, Georgics* and *Aeneid of Virgil*: tr. C. Day Lewis. Oxford Univ. Press, Oxford.

VOLTAIRE. 1947 [1759]. *Candide or Optimism:* tr. John Butt. Penguin, Harmondsworth.

WAITES, Bryan. 1981. 'A newly-discovered letter from Thomas Barker', *Rutland Record* No. 2: 68–9.

WALPOLE, Horace. 1960. *Horace Walpole's Correspondence,* ed. W. S. Lewis, vol. 21. Yale Univ. Press, New Haven.

WARD, W. R. 1958. *Georgian Oxford: University Politics in the Eighteenth Century.* Clarendon Press, Oxford.

WARNER, Sylvia Townsend. 1946. *The Portrait of a Tortoise.* Chatto and Windus, London.

WARTON, Joseph (tr.) 1778 [1753]. *The Works of Virgil,* 4 vols. London.

WEBBER, Ronald. 1968. *The Early Horticulturalists.* David and Charles, Newton Abbot.

WHITE, Gilbert. 1774. 'Of the house martin or martlet', *Phil. Trans. R. Soc. of London* **64**: 196–201.

—— 1775. 'Three letters; of the house swallow—of the swift or black martin— of the sand martin', *Phil. Trans. R. Soc. of London* **65**: 258–76.

—— 1781. 'William Collins, the poet', *The Gentleman's Magazine* **51:** 11–12.

—— 1789. *The Natural History and Antiquities of Selborne.* London.

—— 1795. *A Naturalist's Calendar,* ed. J. Aikin, London.

—— 1911. [MS 1766]. *A Nature Calendar* [Calendar of Flora]: a facsimile, ed. Wilfred Mark Webb. Selborne Society, London.

—— 1975. *Garden Kalendar 1751–1771:* a facsimile, ed. John Clegg. Scolar Press, London.

—— 1986. *The Journals of Gilbert White 1751–1793*, vol. 1, ed. Francesca Greenoak. Century Hutchinson, London.

WHITE, John. 1913 [MS c. 1777]. *An Introduction to Fauna Calpensis,* ed. W. H. Mullens. Selborne Society, London.

WHITE, Thomas. 1785. 'Remarks on oaks' and 'Fraxinus excelsior Linnaei. The ash . . .', *The Gentleman's Magazine* **55**: 109–12, 598–600.

—— 1789. Review of 'Natural History and Antiquities of Selborne . . .', *The Gentleman's Magazine* **59**: 60–3, 144–6.

WILKINS, John. 1675. *Of the Principles and Duties of Natural Religion.* London.

WILLEY, Basil. 1940. *The Eighteenth Century Background: Studies on the Idea of Nature in the Thought of the Period.* Chatto and Windus, London.

WILLSON, E. J. 1961. *James Lee and the Vineyard Nursery.* Hammersmith Local History Group, London.

—— 1982. *West London Nursery Gardens.* Fulham and Hammersmith Historical Society, London.

WILLUGHBY, Francis. 1972 [1676]. *The Ornithology of Francis Willughby:* a facsimile. Paul P. B. Minet, Newport Pagnell.

YARRELL, William. 1837–43. *A History of British Birds,* 3 vols. London.

[YOUNG, Edward] 1728. 'Satire VII' in *Love of Fame.* London.

Index

This index is selective: that is, prominence is given to matters that assist in an understanding of White's (W's) progress as a recorder of the natural scene. Names of species are, in the main, included only if they have been cited in the text or notes in connection with a particular aspect of White's development; further, since White was no systematist (as generally understood), Latin nomenclature has been eschewed.

Abercrombie, John 171n4
    Every Man his own Gardener: annuals in 39; esculents 38; herbs 172n6
accounts, financial (W's)
    'Charity for Schooling' 160n5
    expenses: housekeeping 177n23; proctorship 177n4; during smallpox 165n34; for sister's journey to Lyndon 18
    published 163n18

Banks, Joseph 87–90 *passim*, 133
Barker, Samuel (W's nephew) 143, 197n51
Barker, Thomas (W's brother-in-law)
    communications to Royal Society 165n6, 182n1
    marriage to Anne White 18
    meteorological diaries 18–19
    personal characteristics 19
    visited by W 18, 59
    visits Selborne 56, 194n26
Barrington, Hon. Daines 84
    affability of 91
    annotated two of W's *Journals* 107–8; shown in facsimile (Plate 6)
    introduced W to nat. hist. collections 135
    read W's monography at Royal Society 134–5
    sent W a *Naturalist's Journal* 84–5

    style of writing compared with W's 108
    suggested W prepare a fauna 129, 134; and W's reply 131–2
bat, noctule 201n18
Batty (*also* Battie), Catharine, journal of visit to Selborne 169n29
Berger, Alexander, calendar of 62, 67
birds
    entry of: in *Kalendar* 65–6; in *Journal* 104–5
    prognostic power of 181n36
    Species: aberdavine, in *Selborne* by error 200n9; crag martin (W's 'winter-swallow') 117–19; cuckoo 220; flamingo from Gibraltar: error in calculation of leg-length 218, given to Royal Society 122; flycatcher, behaviour omitted by Linnaeus and Ray 82; mistle thrush 219–20; nightjar: data from Churton 148, dissection 200n8, at Hermitage 170n30, intended monograph 147–8; raven, nesting-tree felled 1; reed-sparrow *see* aberdavine; ring ouzel 202n20; stilt, black-winged 217–18; swallows 191n10, and W's monographies 135; swift, mating of 162n31; warbler, leaf 201n18
blue mist (smog?) 212–13
    entered in *Journal* 105

books (W's)
  Butler's *Sermons* 54
  historical 161n14
  Miller's *Dictionary* 16, 22
  natural history: bought 67; studied
    81–2; sources for 190n41
  obtained from book-club 179n13
  Ray's *Methodus* 22
  recommended to brother at Gibraltar
    122–3
  taken to school 10, 179n15
Bostal 31, 140–1
Bradley, Richard, and salad herbs 172n6
Bristol Hotwells 166n14
  'crisping at', enjoyed by W 21–2
Buffon, on purpose of nat. hist. 202n20

Calendar of Flora see *Flora Selborniensis*
cats, catch birds and bats 187n17
Chandler, Revd Richard 195n33
  assists in study of antiquities 139–40
  considered as addressee for
    'Antiquities' 200n9
Churton, Revd Ralph 144
  advised by W to attend Royal Society
    201n16
  supplied data about nightjar 148
Claridge, John, author of *The Shepherd of
  Banbury's Rules* (weather lore) 182n2
Collins, William, W takes tea with 15
Corpus Christi College, Oxford,
  expulsion of John White (W's
  brother) from 20
Couzens, Alexander, on clouds in
  painting 187n13
crickets 65, 188n29
crops
  damage to, by: birds 13, 162n20,
    162n26; caterpillars and slugs 39;
    dogs 162n20; persons unknown 39;
    weather 38
  harvest of: apple and pear 38;
    cucumber 38
  and light 174n26
  need for improvement 63
Curtis, William, advises on mosses
  197n51

Defoe, Daniel, semi-fictional writing
  201n13
Derham, Revd William, on editing
  letters 157

Drury, Dru, opened specimens from
  Gibraltar 117

eighteenth century
  garden development 25, 29–31,
    170nn31 and 32
  hierarchies in nature 13
  nature in decline during(?) 6–8
  and science: advances 123–4, 202n20;
    letters, periodicals and societies
    201n11
Evelyn, John, *Fumifugium* 212–13
experiments (W's)
  experience of, related to nat. hist.
    176n40
  pursued 51

Flora Selborniensis (1766)
  aesthetic awareness 180n27
  consequences of keeping 77–80
  creatures recorded 74–5
  entries for 1767 in 82
  exotics in, absence of 75
  first publication 178n1
  full title 62
  garden plants in 73
  natural calendar 62–72 *passim*
  nomenclature used 73–4
  pages of, in facsimile (Plates 4, 5)
  prognostic entries 75–6
  raw data for *Selborne* 150
  stages of growth and flowering in 73
flowers *see* plant(s)
Forster, J. R., advises on auroras
  188n27, and insects 190n42
fossils, on Bostal 140–1
frogs, dispersal of 189n34

garden (at Wakes) 23–35
  assistance in 45–6
  boundaries to 174n22
  culinary pleasure from 145
  development of 207–8
  equipment used 47–8
  Heracles statue 32–3
  mount and other structures 30–1, 34
  plants indoors from 171n5
  record of operations in 36–7, 77
  soil in 173n18, 175n35
  sources of seeds and plants 46–7
  transplanting from the wild to 174n29

variety of plants in 37–8, 39–40; *see also* plant(s)
vistas in 30
see also crops, *Kalendar*, tree(s), Wakes
Garden-Kalendar see *Kalendar*
Gentleman's Magazine, The, weather diaries in 97
Gibraltar
  correspondence with John White (brother) 115–28; new role in 115–16
  currents in Straits 125
  fish from, published by Pennant 192n29
  Journal at, by John White 125–6
  public spirit of W in relation to 121–2
  relief of siege (1782) 161n10
  specimens from: first arrived 93–4, 116–22; identification of 117–19, 190n42; preparation for 95; preserved 194n20
  summer migrants, study of 115, 186n43
Gibson, Revd James
  moose, informed W of 184n23
  provides seeds and plants 46
  at Quebec (1759) 4, 166n17
  rector at Bishop's Waltham 22
glow-worm 181n37
gossamer shower 13–14, 22
  and insect life 76
  W's writing-up of 138
Gregorian Calendar 163n1
Griffith, Revd Thomas, at Selborne 186n45
Grimm, Samuel Hieronymus (*Selborne* artist) 138–9

ha-ha
  at Blenheim and Stowe, visited 25
  construction of, at Wakes 24–5
  two such(?), at Wakes 167n5
Hales, Revd Stephen
  preferment 166n10
  status as scientist 165n7
  ventilation, work on 48
  W's appreciation of 20, 158
Hampshire *see* Southampton, County of
Hampton Court
  bridge at, over Thames 160n2
  oak for bridge at 1
  Toy Inn, adjacent to 160n2

harvest mouse
  description of, by Pallas 183n10
  specimen, sent by W to Pennant 81
Henley, Robert (Lord Chancellor), and W's preferment 60
Heracles, statue of *see* garden (at Wakes)
hermitage(s), illustrated (Plate 2)
  built by W 31
  cantaloup feast at 32
  hermit at 31
  inscriptions 169n29
  nightjar at 170n30
Hill, John 70
Hopkins, Gerard Manley, quoted 159, 202n21
Horne, Revd George, at Selborne 169n26
Howard, Luke, and clouds 187n13
Huddesford, Revd George, at Selborne 185n38
Hudson, William, *Flora Anglica* 70
Hunter (at Waverley, Farnham), provides seeds and plants 46
hunting (by W, and W on)
  cost of recovering Fairey 164n20
  decline of deer 164n22
  enthusiasm for 12–13, 21, 56–7
  expenses on 12
  man's propensity for 13
  seminal experience of nature during 13–14
Huxham, John, on coastal climates 98

Inns of Court (Middle Temple), W's father's chamber at 163n10
insects
  in *Flora* 74, 181n37
  from Gibraltar 117
  in *Journal* 104–5

Jenyns, Soame, quoted 164n25
Journal (W's)
  after *Selborne* 146–7, 198n68
  birds and insects in 104–5, 105
  change of use for 106–7
  contextual data 96–101
  entries in: accuracy of 110–14; analysis of and number 95
  familiar animals in, absence of 102–3
  folio from, in facsimile (Plate 6)
  garden records and husbandry in 106
  inspected by Barrington 107–8

Journal (W's)—*continued*
  miscellaneous observations 105–6
  plant record in 103–4
  rainfall and wind record 97–8, 100
  temperature and pressure record
    98–100
  weather record 100–1, 105–6
  see also *Naturalist's Journal*

Kalendar (W's)
  dates of annual commencement 165n5
  decisiveness of entries 37
  duration of 167n1
  first entry 18
  folio from, in facsimile (Plate 3)
  importance of, in W's development
    48–9
  influence on: by Thomas Barker
    19–20; by Stephen Hales 20
  loss of focus in 79–80
  memoranda in 174n25
  narrative reward of 37, 40–4
  natural history in 52–3, 65–7
  raw data for *Selborne* 46, 150
  status of, consequent upon *Flora* 77–9
  technical terminology in 175n34
  weather records in 77–9
  see also measurement *and under* records
Keats, John, quoted 159, 202n21
Knight, James, ponds of 171n3
Kuckalm (T. S. Kuckhan) 217

lacewing, from Gibraltar 120–1
Lee, James, book by, sent to Gibraltar
  124
Lightfoot, Revd John, letters from, and
  Selborne visitor 133, 186n45,
  194n14
Linnaeus, Carl 63, 109, 180n23

Magdalen College, Oxford, manorial
  court of, at Selborne 169n26
Mander, Revd Thomas, friend of W
  165n35
Marsham, Robert 160n6
  congratulates W on *Selborne* 148
  prompts W to measure trees 2
Mawe, Thomas *see* Abercrombie, John
measurement
  homely devices for 50–1
  progress towards standard scales
    187n11

by W of fauna 82
see also under *Journal* (weather
  record)
meteorology
  comparison of climates proposed
    125–6
  diaries of, at Selborne 97
  folk guidance concerning 182n2
  letters about, in *Selborne* 99
  W's: delight in 188n27; inclusion of,
    in Gibraltar history 125–6;
    narrative of 77–9
  see also under Barker, Thomas; blue
    mist; *Journal*; *Kalendar*
migration, species separation 188n26
Milton, John, quoted 33
Montagu, George, congratulates W
  199n71
moose 184n23
Mulso, Revd John
  claimed W had arrived at
    independence 54
  distaste for 'out-door work' 179n14
  encouraged W towards marriage 60
  opinion that *Selborne* would
    immortalise both author and
    subject 138
  shared literary reading with W 15
  sought poetic inspiration 16
  W's work on antiquities
    'mal-apropos' 140; *see also* 195n34
Musgrave, Revd Chardin 58, 185n37

natural calendar
  and artificial calendars 64
  birds in 75–6
  and *Flora* 72–6
  guide to husbandry 63
  outcome of, for gardeners 64; in
    prognostic plants 64, 71–2
  W sceptical of search for 108–9
  see also Berger, Alexander;
    Stillingfleet, Benjamin
natural history
  approach to study and writing of
    122–28
  collection of data by 'Queries' 81
  disparate data of 111, 126–7
  indeterminacy in 189n38
  and Robert Plot 155
  study of, undifferentiated 123–4
  see also *Flora Selborniensis*; *Kalendar*

Natural History of Selborne see *Selborne*
Naturalist's Journal (for W's use of, see
    Journal)
  given to W 84
  imprint for 184n19
  page design in 85
  public purpose 85–6
  W's purchase of 184n19

Oriel College, Oxford
  elections at 55–6
  preferments offered to W 178n26
  W admitted to 10; elected fellow 15;
    nominated proctor by 20;
    candidate for provostship at 57–8
otter, killed at Selborne 151
Oxford University
  church and state at 55
  college fellowships at 58–9
  education at 164n30
  election of chancellor (1759) 178n28
  gardens at 167n8

Pennant, Thomas 80
  proposed visit to Goodwood 86, 88
  'Queries' of 81, 214–16
  recognition of W's knowledge 194n23
  W's opinion of 192nn18 and 29
plant(s)
  sources for 46–7
  species of, in *Kalendar* 37–40, 68–72
  Species: amaranths and cockscombs
    40–2; *Arundo donax* 190n40; asters,
    China, example of W's access to
    new forms 173n17; butcher's
    broom 117; crown-imperials
    175n33; cucumber and melon 45;
    mushrooms, from spawn 45;
    orchids, transplanted 187n21;
    potatoes: associated with Popery
    173n21, culture of 42–4, introduced
    at Selborne 44, slow acceptance of
    173n20; sea-kale 190n40; toothwort,
    for Joseph Banks 87–8
Plot, Robert
  Natural History of Oxford-shire 155
  originator of county natural histories
    158
poetry (W's)
  'A Harvest-Scene' 198n56
  'The Invitation to Selborne' 16, 28–9,
    60–1, 209–11

'On the Rainbow' 198n56
  translations by 16, 34, 54
Pope, Alexander
  on gardens 30
  quoted 3
Providence
  balance of sexes owing to 7–8
  and intricacies of nature 67
  W's faith in 151, 202n21

rabbit warrens 187n18
Ray, John
  candour of 192n18
  Methodus, used by W 22
  Philosophical Letters of, used by W
    156–7
  quoted for sexual balance of species
    7–8
  Synopsis (3rd edn), used by W 70
records (W's)
  before *Journal*, private 84–5
  botanical, in *Kalendar* 67–72
  brewing 182n3
  continuities in 75–6
  corrections to 52
  elegance of hand in 52, 72
  enthusiasm in 76
  reliability of 51–2, 83–4, 110–14,
    190n39
  weather: see under *Flora Selborniensis*;
    Journal; *Kalendar*
Richardson (of Bramshott), visited by
    W 198n57
Richardson, Samuel 15, 157
Royal Society
  exchange of information through 156
  receives W's monographies 135
  weather records for 182n1
  W's gift to 122

Scopoli, Giovanni 118
  Annus of 191n12, 193n5
Selborne
  botanical variety at 83
  context for child-rearing 34–5
  cottages in 162n24
  in county of Southampton 161n9
  described to entice Banks to visit 89
  historical buildings at 168n19
  'The Invitation to Selborne' *see under*
    poetry

Selborne—*continued*
  loss of oak on Plestor 1 (*see also* Storm,
    Great)
  population 7–8
  scenery of, 'enlarges' 27–8
  topography 27
  W's consort 60–1
  W's extended garden 27–35 *passim*;
    illustrated (Plate 1); map of 206
Selborne, Natural History of
  'Advertisement' to 158
  artist for 138–9
  beginnings of, in Barrington's interest
    129–30, 134–5
  completion of, delayed 138–45
  to comprise letters and *Annus* 135
  considered as a collaborative volume
    156
  copy text for 151, 152
  dating of letters in 185n40
  encouragement towards, by Lightfoot
    and Skinner 133
  form of, importance of 157–9
  indexing of 146
  introductory letters in 155–6
  letter form, use of 156–7
  Letter 29 (to Pennant), analysis of 151
  mottoes for 34, 149
  prima facie judgement of, questioned
    151
  progress towards, in Gib. letters 115,
    122–8
  prospectus for 137
  publication and full title 146, 160n1
  reviews of 146
Sheffield, Revd William
  account of Banks's specimens 123
  at Selborne 92–3
  too busy to help 194n11
  understanding of formal description
    superior to W's 118–19
Skinner, Revd Richard
  exhorts W to write a nat. hist. 133
  laziness of 194n11
  at Selborne 93, 124
  in Wales 194n13
Solander, Daniel, met by W 89
soldiers, 'winter quarters' 183n14
Southampton, County of 161n9
Stillingfleet, Benjamin
  Calendar of Flora, kept at Marsham's
    148

  doubts value of astronomical
    calendar 71
  Miscellaneous Tracts 62–5, 85; calendar
    recommended to John White at
    Gib. 126; prognostic value of birds
    75–6; studied 67
  see also natural calendar
Storm, Great (1703) 1, 160n4
Stuart, Sir Simeon
  pineapples of 38
  provided 'Indian-turnep-seeds' 46
  superstitions, deplored by W 182n4
Swift, Jonathan, quoted 6, 162n28

Thomson, James, *The Seasons* 10
Timothy (W's tortoise) 197n47
tree(s)
  despoliation of, at Selborne 27
  as groves for religious purposes 27
  loss of, in vistas 169n25
  scattering of seed for 168n12
  timber from 1, haulage of 162n23
  at Wakes, planted by W 2
  Species: beech: W admires 1,
    instructs mast from, to be scattered
    168n12; elm, planted by W's
    grandfather 1; oak: for bridge at
    Hampton Court (*q.v.*) 1, largest in
    Britain near Selborne 199n72,
    measured by Marsham 2, for navy
    1, 5, nested in by ravens 1, on
    Plestor 1
truffles
  in Wiltshire 56
  W's records of 198n60

Virgil, *Georgics* of, quoted by W 2, 29
Voltaire, *Candide*, quoted by John
  Mulso 4

Wakes (W's home)
  building of parlour at 141–2
  family home 143
  garden laid out at 10
  history of 163n2
  naturalists at 93
  notable trees at 1
  well at, depth of 162n21; cleaned by
    W 163n6
  see also garden (at Wakes)
Walpole, Horace, quoted in connection
  with threat of French invasion
    161n15

Warton (Joseph and Thomas) 10
White family
  and Barkers of Lyndon 18
  contributions to nat. hist. 182n5
  genealogy 204–5
  pattern of inheritance in 170n34
  as visitors at Wakes 142–5, 178n31
White, Benjamin (W's brother) 18,
    165n3, 182n7
  Fleet Street shop of 80
  publisher of: later editions of *Journal*
    184n19; *Selborne* 18
White, Revd Charles (W's uncle) 60
  accepts W as curate 16
  provides plants for Wakes 46
White, Revd Gilbert (W's grandfather)
    2, 11
  father (Sir Sampson White) 11
  memorial tablet to 1–2
  portrait 163n14
  services to Selborne 2, 9, 10
White, Mrs Gilbert (née Rebecca
    Luckin, W's grandmother) 11, 56
White, Revd Gilbert (author of *Selborne*)
  Chronology of events during his life:
    birth 9; childhood and schooling
    9–10; university 10; death of
    mother 10; parental influences
    10–11; life in Oxford 12–14;
    graduation 15; literary reading
    15–16; ordination and first curacy
    16; begins *Kalendar* 18; proctorship
    20–1; curacies at Durley, Dene and
    elsewhere 22, 54, 56–7, 58, 60;
    prospect of marriage 34–5, 58–9;
    candidate for provostship 57–8;
    death of father 58; begins *Journal*
    84; role as head of family 115,
    143–4; at height of his powers
    119–20; brother (John White) at
    Wakes 136–7; builds parlour
    141–2; Mrs John White at Wakes
    144–6; publication of *Selborne* (q.v.)
    146; final years 146–9; *see also* Wakes
  Character and attitudes: admiration
    of Banks's temerity 89–90; advice
    on clerical behaviour 197n53;
    appreciation of paintings 164n27;
    apprehension of social unrest
    201n16; candour 129, 192n18;
    delight in 'natural knowledge' 91;
    enjoyment of the 'irregular' 27, 28;

  faith in a beneficent Providence
    202n21; fastidious for church
    cleanliness 165n8; lacking qualities
    necessary in a provost 57–8;
    literary decorum 176n1; military
    interest 3–4, 161n10; mutuality of
    confidence with brother (John)
    136–8; procrastination 116;
    propriety 103; pride in nephews
    and nieces 197n55; restlessness,
    and life at Oxford 16–17; *see also*
    accounts; books; hunting; poetry
  Excursions and tours: Bristol
    Hotwells (q.v.); Devon 16–17;
    Lyndon, Rutland 14, 18, 20–1, 59;
    other places 14, 16, 56, 166n11,
    167n9
  Naturalist and writer: accuracy
    110–14, 120; aim of scientific
    endeavour 7, 90; astronomy
    188n27; began correspondence:
    with Barrington 92, with Pennant
    81; behavioural observations 81–2;
    born in his garden 48–9;
    correspondence compared 92;
    development as, summary of 90,
    119–20, 158; discoveries 201n18,
    202n20; dissection of specimens 82;
    first known records 18; ichthyology
    83, 102; invited to collaborate, by
    Barrington 189n31; *Journal* received
    from Barrington 84; limitations in
    formal description 118–20; local
    studies 129–30, 193n5; met Joseph
    Banks 87–8; persistence of 120–1;
    Selborne (q.v.): editing of 151–6, 157,
    inaccuracy in 217–21; skill as
    observer and record-maker 112–14;
    spontaneity of records 65; see also
    under *Flora Selborniensis*; *Journal*;
    Kalendar; meteorology; records
  Botanical studies: authorities used 70;
    beginnings in 67–8, 179n21, textual
    problem of 179n17, no more than a
    bauble! 179n12; differentiation of a
    flora from a hortus 72–3; plant
    records of, in *Kalendar* 68–72,
    locations given in 68–9,
    nomenclature used in 69–70, stages
    of growth enumerated in 71–2;
    reasons for 67, 69; sought seeds
    from Pennant 83

White, Revd Henry (W's brother)
  as hermit 31
  journals of 146
White, John (W's father) 9, 10
  character 10–11, 163n9
  decline and death 57–8 *passim*;
    communicated to John White (W's
    brother) 115
  interest in music, and the weather 11
  testamentary provisions 58
White, Mrs John (née Anne Holt, W's
  mother) 9, 10
White, Revd John (W's brother)
  constructs Zigzag 21
  expelled from Corpus Christi, Oxford
    17, 166n9
  'Fauna Calpensis' 192n30
  at Gibraltar 4
  recipient of W's first thoughts about
    Selborne 135–8
  at Selborne 136
  sought return to England 115
White, Mrs John (née Barbara
  Freeman, W's sister-in-law) 93, 145
  frees W to complete *Selborne* 145–6
  keeps house at Wakes 136–7, 145–6
  visits England (1769) 93, 115

White, John (W's nephew, 'Gibraltar
  Jack')
  arrives in England 115
  'comptroller general' of works at
    Wakes 141
  description of fructification
    applauded 124
  education supervised by W 93, 115
  hand in Selborne MS 200n6
  letter by, facsimile of (Plate 7)
White, Mary (W's niece), assists W
  143–4
White, Richard (W's nephew), receives
  instruction from W 143
White, Thomas (W's brother)
  arboriculturist 199n72
  inheritor of Holt monies 196n38
  interest, in antiquities 195n36; in
    botany 82, 182n5
  provides plants for Wakes 46–7

Yalden family, how related to W's
  family 196n40
Yalden, Revd Edmund, offered curacy
  to W 56
Young, Edward, quoted 3, 161n8

Zigzag, construction of 21